Rheinisch-Westfälische Akademie der Wissenschaften

Natur-, Ingenieur- und Wirtschaftswissenschaften   Vorträge · N 238

Herausgegeben von der
Rheinisch-Westfälischen Akademie der Wissenschaften

VICTOR POTTER BOND

The Impact of Nuclear Power
on the Public:
The American Experience

Westdeutscher Verlag

Sonder-Vortragsveranstaltung der Klasse für
Natur-, Ingenieur- und Wirtschaftswissenschaften
in der Kernforschungsanlage Jülich
am 24. Januar 1973

ISBN 978-3-531-08238-7   ISBN 978-3-322-88501-2 (eBook)
DOI 10.1007/978-3-322-88501-2

© 1974 by Westdeutscher Verlag GmbH Opladen
Gesamtherstellung: Westdeutscher Verlag GmbH

Contents

Summary of the Development of the Controversy	8
Detailed Review of the Controversy	10
The Need for more Power	13
Availability of some Natural Energy Resources	15
Different Approaches to the Problem of Power	17
Severe Restriction of Power Consumption	17
Nuclear Power	18
Effects of Radiation	20
Severe Accident	24
Waste heat	31
Long-term Waste Disposal	32
Transportation of Wastes	33
Fuel reprocessing Plants	34
A Comparison of Reactor Types	34
Plutonium Toxicity	37
Risks of alternative Sources of Power	38
Concluding Remarks	43
Summary	45
Zusammenfassung	47
References	49
Tables	51

The Impact of Nuclear Power on the Public: The American Experience

Von *Victor Potter Bond*, Associate Director
Brookhaven National Laboratory, Upton, New York

It is useful to examine the so-called "nuclear controversy" in the light of American experience, since it is here that it has been manifest in the protracted debates by scientists, laymen and the press over the pros and cons of this relatively new source of energy. I shall attempt, in the short time available, to review this experience and indicate the direction in which the overall course seems to be moving. In so doing, I wish to recount briefly, in approximately chronological order, the events as they have occurred and are ocurring. Although the specific issues have changed, the "controversy" is by no means over in the United States, and it is difficult at this point to predict with any degree of confidence what course will eventually be pursued.

When someone speaks or writes on a controversial subject, the audience has the right to know something of the individual's background and qualifications so that they can make an assessment of his overall attitudes and perhaps biases, as an aid in evaluating what he has to say. I am primarily a physician with a background also in physics and medical physics. My scientific career has been concerned largely with the biomedical effects of various types of radiations, and I have been closely associated with studies on the effects of the radiations produced in nuclear processes. I have taken part in the "nuclear controversy" in the United States, principally in the context of the potential effects of radiation from power reactors, and have endeavoured to counteract some of the more extreme statements that have been made by some critics of nuclear power. I am employed by the Brookhaven National Laboratory, which is operated for the U. S. Atomic Energy Commission by a private corporation made up of nine of the leading universities located in the Eastern United States. What I say here represents my own personal viewpoints, and does not necessarily represent those of the Brookhaven National Laboratory, The Associated Universities, Inc., the Atomic Energy Commission, or any other governmental agency or organized group.

It should also be made clear that when I speak of radiation effects on biological systems including man, I speak from extensive direct knowledge and experience in this area. However, in my discussions of more engineering-related subjects, e.g., possible severe accidents and waste storage, my knowledge is derived principally from extensive reading, and discussions with those highly experienced and knowledgeable in these fields.

It is obvious that the ultimate choice as to how much (if any) additional power will be provided (and in what form, fossil, nuclear or otherwise) cannot be determined solely by the government, by power companies, by scientists or by any other groups. Public acceptance of any course is mandatory, and it is ultimately the people as a whole who will determine public policy and programs.

I believe that nuclear power should and must play a markedly increased role in providing electrical power. However, I am not here to "sell you" on any particular solution to the overall problem. My objective is to give you the background information to aid you in understanding the American debate and to help you to arrive at an informed opinion with respect to the direction you would like power development to proceed.

Summary of the Development of the Controversy

The "nuclear controversy" has undergone considerable evolution in the United States over the past several years. The essence of the developments as I see them are as follows: Initially, both critics and defenders focused almost exclusively on one form of power, nuclear, to the almost total exclusion of consideration of either the benefits or the risks of other possible types of power. The critics centered almost exclusively on the problem of potential hazards of radiations from these nuclear devices, and the proponents responded accordingly. The tone of the presentations and debate was often strident, with extreme positions being adopted by both sides. This led in general to much more heat than public enlightenment.

As awareness of an impending energy crisis has developed, the arguments have assumed a larger frame of reference. Thus both groups have tended to take more moderate positions, and have also attempted to consider other sources of power, the possible disadvantages and hazards associated with these sources, and the advantages, disadvantages and possible hazards of a number of alternative sources. At the same time the Atomic Energy Commission has made some changes that have tended to allay the fears of some of the critics of nuclear power. Former Chairman Schlesinger made it clear that the Atomic Energy Commission (AEC) is not a partner of the power companies and is not attempting to promote nuclear power (with the exception of the breeder reactor, as the result of a Presidential directive). Also, much more restrictive routine release rates for radioactivity have been proposed by the AEC. Although one can justify these more restrictive release rates on the basis that they are technologically possible and that all exposures should be kept as low as practicable, many believe, on the basis of cost-benefit analyses, that the proposed release rates are excessively and unnecessarily restrictive [1] and that the change came about in large measure as a result of extreme pressure from the critics.

As a result of the famous "Calvert Cliffs" decision, the Atomic Energy Commission was forced to alter its regulatory practices to achieve compliance with the National Environmental Protection Agency (NEPA) regulations, and now, as a result, extensive "impact statements" are required for each new reactor.

With the fading of radiation risks from routine reactor emissions as a prime focus for critics, other issues, principally that of reactor safety and particularly the adequacy of the emergency core cooling system (ECCS) have come to the fore. A number of articles pro and con have appeared on the subject, and public hearings are currently being held.

Recently Ralph Nader, after consultation with several nuclear critics including John Gofman, Ernest Sternglass, Daniel Ford and James Mackenzie, has entered actively into the scene and is demanding a moratorium on the construction of nuclear power plants. Although many believe he has entered too late and covers ground already thoroughly plowed, it is difficult to predict where this move will lead.

On the other hand, the press and the public have become increasingly aware of the extent of air pollution and the resultant deleterious effects produced by the burning of fossil fuel. Articles on the adverse effect of this on individuals, particularly those with lung disease have appeared. Damage to plant life is obvious in the Los Angeles area, as in other countries as well, e. g., Sweden and Germany. Gas masks have been ordered by the Labor Office in Venice [2]. Severe damage to the coliseum in Rome, to the Cologne Cathedral [3] and buildings in Sweden [4] is ascribed to sulphuric acid produced when sulphur oxides are released into the air with the burning of fossil fuels.

Also, serious investigations of the possibility of "cleaning up" and making more efficient all sources of energy now available are being made. Gasification of coal is receiving increasing attention, having been identified as a priority goal by President Nixon (Clean Energy Message to Congress, 4 June 1971). Such approaches as combined gas and steam turbines which may appreciably increase the efficiency of fuel consumption, are also under consideration. It is increasingly realized that no one source of power is adequate, that there is a proper role for many available sources and that all need additional development work.

The American public is also increasingly aware of the limited availability of fossil fuel resources, not only because of innate scarcity but also because some abundant fuels, e. g., coal, simply are not produced in adequate amounts in the U.S.A. Natural gas is in very short supply. Some 90 % of the oil for power plants on the eastern seaboard is imported. The implications of the dependence of these vital supplies on international political considerations are increasingly realized, as is the severe hazard associated with a significant lack of power for even a brief interval. Articles on real, not predicted energy shortages appear almost daily in the American press.

More and more individuals and groups are now attempting to look objectively at the overall advantages and risks of nuclear and other forms of power to form a basis for choice in a given situation. A large number of rather extensive studies along these lines are now being conducted. Data are incomplete as yet; however, considerable progress has been made. It is hoped that over the next few years, these studies will have progressed to the point that one is in a position to make decisions based primarily on logic, rather than on emotion and often badly informed public opinion. Thus one can determine rationally the role that each potential form of energy should play in providing the needs of an increasingly energy-voracious population.

The recent and real energy shortages in the U.S.A., that required the closing of some schools and factories, have made "converts" of many Americans [*]. Public opinion polls show that most have and do favor more nuclear power.

Detailed Review of the Controversy

Now let us review, in a little more detail, the development of the "controversy" in the United States. This controversy (over the potential hazards from the peaceful atom) began as early as the late 40's when at least one individual, Leo Goodman, a labor union official located at Oak Ridge, Tennessee, began to raise his voice that the AEC was not taking proper

---

[*] The Governor of Minnesota has recently (Nucleonics Week, March 1973) reversed his decision on a moratorium on more nuclear power plants in that state, on the basis of severe energy shortages and because there is now as much discussion on the environmental effects of coal-fired plants as there is about nuclear plants.

cognizance of what he claimed were radiation-induced injuries to workers in its facilities. The origins to some degree go further back, and are associated with the use of atomic weapons in Hiroshima and Nagasaki. In the mind of the American public nuclear energy is still equated in large degree to atomic bombs. While this has no basis in fact, it has been capitalized on frequently by critics of nuclear power. Several others joined in these still relatively low key criticisms of the Commission's awareness of radiation hazards, e. g., during the early attempts of the New York Con Edison Company to construct a nuclear plant at the Ravenswood site, and the bitterly criticized Enrico Fermi plant near Detroit.

These initial forays were essentially all at the instigation of individuals, and environmental protectionist groups were involved only to a very limited extent. Several additional individuals joined the opposition, most notable among them David Lilienthal, the first chairman of the AEC, from 1946 to 1950.

The problem of testing nuclear weapons above ground played a large role in developing the overall public attitude towards radiation, and led to the opposition of the Nobel Prize winner, Linus Pauling, who in 1960 produced his first antinuclear book, "Life or Death in the Nuclear Age". Thus the opposition to nuclear power was intertwined with opposition to nuclear weapons and their use, and was complicated by a growing distrust of politicians, the government and big business.

Although opposition did result in action with respect to above ground nuclear testing and its associated fallout, these activities had little effect on the initial development of nuclear power.

The first large-scale opposition by anti-nuclear groups took place in connection with the second Indian Point reactor in New York, in 1966. By this time a number of organized groups had emerged. These included the Friends of the Hudson and Conservation Center, the Save New England, Inc., the Pollution League, and the Council for Education and Sciences, etc. At this time additional books, including those by Sheldon Novick (The Careless Atom) and Elizabeth Hogan (Perils of the Peaceful Atom) helped to increase public concern and unrest.

Perhaps the controversy can be said to have started in earnest when Ernest Sternglass, and Gofman and Tamplin appeared on the scene in the late 60's, and I shall discuss their claims later. The principal reason for the significant impact of the statements of these individuals, as opposed to most previous critics, was that they were scientists and professors, with an impressive list of credentials, and experience with the uses and effects of radiation. In addition, they are excellent public speakers who, with apparently inexhaustible energy, debated and testified frequently and at length, over the entire country.

It was at about this time that additional potential hazards associated with nuclear energy received some attention, although in vague and nonspecific terms. These included so-called "thermal pollution", reactor safety and accidents, transportation of radioactive materials, reprocessing plants, and the long-term storage of high level radioactive waste.

A significant factor in the development of concern on the part of the public was the enormous coverage given in the press to the very vocal critics. Almost daily, in some of our most influential news media, one encountered such statements or headlines as "32,000 additional cancer deaths", and "scientist predicts enormous increases in infant mortality from radiation exposure".

A significant milestone, covered widely in the press, was the controversy with respect to the licensing of the Monticello plant, Northern States Power Company, Minnesota. The issue concerned "states rights", i. e., whether the federal government had the exclusive right to regulate emission standards from reactors, or whether an individual state could impose even more restrictive standards if it so wished. Litigation finally ended in a ruling favorable to the federal government. Organized opposition groups appeared at the hearings for the Monticello reactor, and this triggered the proliferation of such groups all over the country. Often two or more such groups were active at the same time in a particular case. A national coalition of opposition groups has appeared, which combines the resources and talents of the individual groups.

More recently, following the AEC proposal to reduce reactor effluent limits, attention has been focused mainly on reactor safety. Perhaps the key article in this respect was that by Ian A. Forbes, et. al., of the Union of Concerned Scientists (UCS), which appeared in the September 1971 issue of Nuclear News [5]. This article stressed the extreme dangers to which the authors consider the American Public may be exposed as the result of a hypothetical "catastrophic" accident in a nuclear power plant. In many respects, this situation is similar to the earlier one involving Dr. Sternglass and Drs. Gofman and Tamplin. The report was written by a group of Boston area professors, which included invidivuals with impressive credentials. This has been followed by additional articles by the same group, and I shall discuss later some of their allegations and the views of those who disagree.

By this time radiations from routine releases of reactors appear to have become almost a "dead issue" as far as the media and the public was concerned, and the issue of nuclear safety, particularly the emergency core cooling system (ECCS) emerged to the fore. Public hearings on this issue, held in Bethesda, Maryland, have just been concluded; however, it will probably be several months before the Commissioners can digest the voluminous record and issue a decision on whether and to what extent a revision of regulations is justified.

It should be stressed that even though there exists a considerable number of antinuclear individuals and groups, they represent in their entirety only a very small percentage of the total population and thus a minority view. The "hard core" opponents number perhaps in the few hundreds; the total, many of whom seem motivated by the wish to be associated with what they regard as a "good cause", perhaps represent no more than a few thousand. Very few of the anti-nuclear groups advocate an abolition of nuclear power, although some individuals and groups would of course like to do this. Most seek mainly to ensure that the potential hazards of nuclear power have in fact received adequate attention and that the community at large is being protected to the maximum degree. For example, the Sierra Club, a large and influential group which has been very vocal in its opposition to some of the practices associated with current nuclear reactors, has stated that nuclear power plants, for most situations, appear to offer the best source of electrical power for the immediate future, and that minimum environmental damage is associated with this form of energy. They do not oppose nuclear power plants in principle.

The Need for more Power

Figure 1 depicts the rate of increase in world population, which in itself implies a rapidly increasing need for power [6,7]. Figure 2 shows the per capita rate of energy usage in countries as a function of per capita income. Figure 3 shows rate of increase of *per capita* use of electrical power in the USA. These factors lead to the rather startling projections of the increased need for electricity in the USA shown in Fig. 3; with the requirement doubling in less than 10 years. The demand for additional power in developing countries is real and urgent. Table I shows the sources of energy utilized in the USA in 1971 for all purposes. It is of interest that the *daily* energy need of the average American household requires the use of the equivalent of approximately 21 kg of coal, 4 liters oil-equivalent of hydropower, 9 liters of oil products, 25 liters of natural gas and ½ liter oil-equivalent of nuclear power [8].

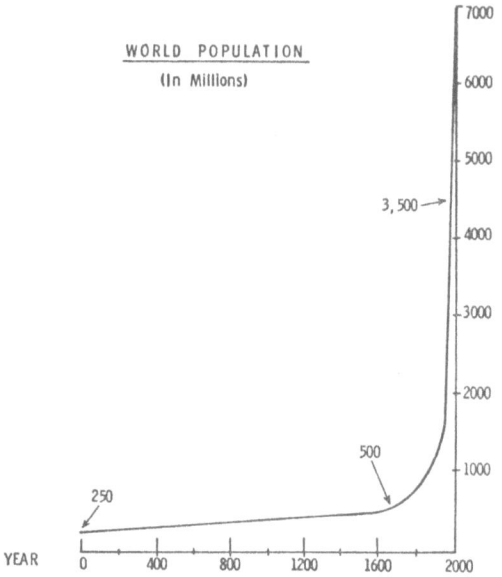

Fig. 1: Rate of growth in the world population over the past 2000 years [6, 7]

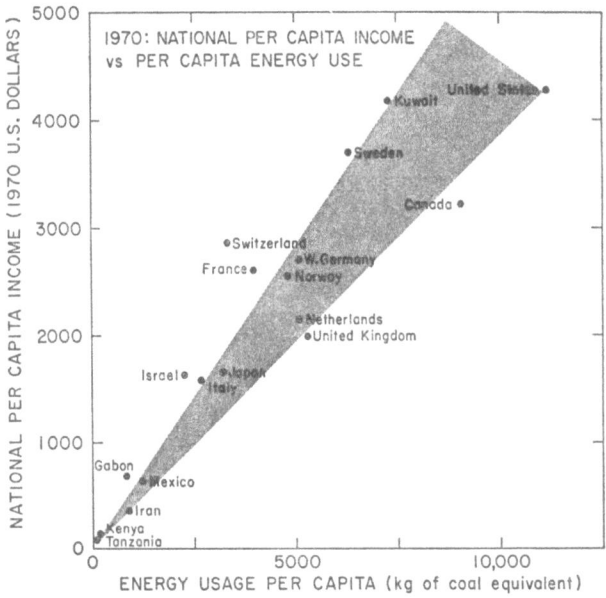

Fig. 2: The energy uses per capita versus the national per capita income, for different countries [6, 7]

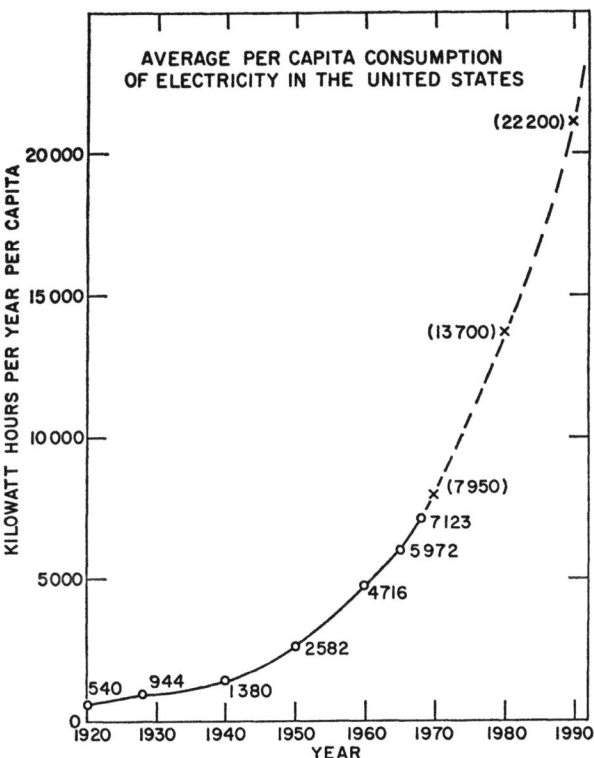

Fig. 3: Average per capita consumption of electricity in the United States for the past few decades and projections of consumption for the future [6, 7]

Availability of some Natural Energy Resources

Most natural energy resources will be almost completely depleted, at the present rates of consumption, within limited periods of time. Figure 4 shows schematically the rate of production after the time at which the resource is first used for energy production. The inverse of the curve, particularly at later times, probably reflects rather well the relative cost per unit amount.

It is difficult to predict accurately how much longer certain fuels will last, since this depends on a number of factors including the amount of time and money expended in searching for new resources, environmental considerations, the degree to which foreign imports are utilized, and the amount of money the public is willing to spend for the fuel as availability declines and/or one is forced to use poorer-grade ores or other materials. Approximations, however, can be made. Natural gas in the USA is on the declining part of the curve in Figure 4, and the available resources will last only a few more years, probably no more than a decade. Oil similarly is on the declining portion and will probably last no more than one or two decades (oil resources off the eastern seacoast can be tapped if the objections of environmentalists are ignored; however, the available total is estimated at some 5.5 billion barrels, less than one year's supply for the entire USA). More costly shale oil could last on the order of 30 to over 100 years, depending on the cost public will accept. Coal will last for several centuries, perhaps four or five, depending again on the tolerated cost in terms of dollars and environmental damage. The availability of uranium suitable for use in thermal (non-breeder) reactors is measured in decades to perhaps a century, depending again on cost considerations; with breeder reactors these fuels will last several centuries.

The following obvious point should be stressed:
Almost all of the above fuel materials are useful for many purposes other than for "central station" heating plants, e. g., oil and coal are used for motor, rail and air transport, important chemical products, etc. Thus the urgency to develop non-fossil fuel alternatives for central station power plants is greater than indicated by the above time figures.

Different Approaches to the Problem of Power

When I speak here of power or additional power, I refer to central station generating plants that provide electrical power for most of its common uses for every day work and living. I exclude the myriad additional requirements and sources of power such as gasoline for automobiles, fuel for aircraft and trains, etc., although these have to be taken into account in any overall assessment of the energy requirements of a society. Among the many possible approaches to providing additional power are those listed in Table II.

In discussing these possible sources of additional power, I should like to first eliminate wind, tidal action and harnessing ocean currents on the grounds that their practicability is highly questionable and/or undemonstrated, and that the total recuperable energy would be insignificant in the overall picture. Also, while I certainly do not wish to dismiss controlled thermonuclear reaction plants, we have no assurance if and when this approach will be feasible. Thus, while a great deal of attention must be given to the possible use of fusion reactors, it does not appear sensible to orient our present plans around this source of energy. Similar comments apply to

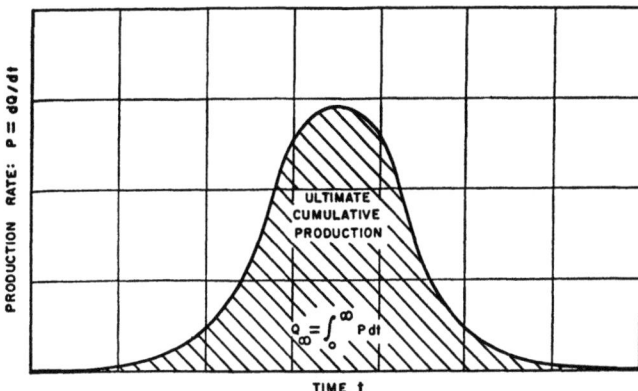

Fig. 4: Schematic diagram showing the production rate of fossil fuel energy sources as a function of time after their initial discovery. The cost of the fuel would be inversely proportional to the production rate, approximately [6,7]

solar energy. Geothermal energy in its present form is available in limited regions, and certainly this mode should be developed. However, its application on a wide scale would require new technologies just now in the early stages of speculation. Additional hydroelectric power is in limited availability and is opposed by environmental groups even more strongly than some oppose nuclear power.

Let me then turn to the approaches that can be considered feasible at the present time, namely little or no additional power, or such readily available sources as natural gas, oil, coal and nuclear.

Severe Restriction of Power Consumption

As previously indicated (Fig. 3), power consumption has been increasing at a high rate in the United States, with a doubling time of ten years or less [6,7], and most projections of power needs for the future have been obtained by extrapolation from this base. Critics of nuclear energy have pointed out that power consumption is rising at a much faster rate than is the population and have stated that this is due in some measure to advertising by the utility companies to increase power consumption and thus their profits. Some more extreme critics have advocated zero power growth, arguing that additional power will be used for such "essentials" as aluminum beer cans and aluminum trays for TV dinners. There has been increasingly serious attention given to attempting to reduce the waste in power consumption, and thus to reduce the total need for energy [9]. The proponents point out that much energy is now consumed in an unnecessary or wasteful manner, and that the demand for energy growth is partly a matter of choice. One must examine just what standard of living one wishes to enjoy, and must evaluate its compatibility with maintaining the overall quality of our natural environment. There is much merit in such analyses and they deserve serious consideration.

However, even strong advocates of reducing the rate of energy consumption recognize that the rate of power consumption will continue to increase at an appreciable rate, although hopefully at a slower rate than predicted earlier. This is true of countries that now enjoy relatively high standards of living, and certainly is and will be true for developing countries.

Therefore, one must work from the assumption that most of the predicted increased power needs have to be dealt with on both a short term and a long term basis. It must also be recognized that any source of additional power, no matter what this may be, carries with it hazards to human beings and to the environment. Those who adopt the attitude that any risk to man or his environment is unacceptable are simply not living in the real world.

Thus as I see it, the task before society is to look objectively and dispassionately at the potential energy sources that are available over a given period of time and to make as good an assessment as is possible with respect to the benefits and costs associated with each type of power production in a given locality and situation. No one potential power source should be considered or examined to the exclusion of others. Only when society is armed with such information is it in a position to make a rational decision with respect to what is best for it and its environment. It would be sad indeed to have any given type of power production rejected and its potential benefits ignored because of irrational fears of some particular aspect of real or supposed risks associated with that form of power.

Nuclear Power

I strongly believe that all approaches to filling additional power needs should be used and developed since no one form is likely to solve all of the problems associated with power development. From my analyses, I believe that nuclear power should and will be employed to supply a large portion of these needs. However, I certainly am not attempting to sell you nuclear power for any particular situation. What you wish to do is of course your decision.

Some of the advantages of nuclear power, which I shall elaborate on, are that it is clean, that it is economically competitive, that it is as safe if not safer than most other forms of power production, and that the supply of fuel is essentially inexhaustable if breeder reactors are used. The disadvantages are principally that, at the same power rating for some contemporary (thermal) reactors, the amount of thermal waste is greater than with fossil fuel. However, with present and future high temperature gas cooled (HTGR) and breeder designs, nuclear is and will be at least as efficient as most fossil fuel cycles in this respect.

The second potential disadvantage of nuclear power has to do with the possible hazards associated with fission products. These must be examined in the context of several circumstances related to the overall cycle of obtaining, using and disposing of nuclear fuel. These include 1) mining, 2) routine releases, 3) severe accidents, 4) transport of radioactive material, 5) fuel reprocessing, and 6) long term storage of high level radioactive wastes. I shall deal with these in turn.

All mining operations are notoriously hazardous with an accident rate higher than for any other major industry. Also, workers in mines containing appreciable amounts of uranium and its daughters (many different types of mines; not only uranium mines), are exposed to a risk of lung cancer, particularly from radon and its daughter products. Although there is no question that there is an increase in lung cancer rate among these miners, the quantitative aspects are not clear. The situation is severely compounded in that many miners are or have been heavy cigarette smokers, which in itself is of course capable of producing lung cancer.

One can say, however, that the total cost from nuclear fuel mining is considerably less than with deep coal mining in which "black lung disease" is highly prevalent. This is mainly because the total amount of mining required for a given level of energy production is considerably less for nuclear power than for coal-fired plants. The annual operation of a 1,000 MWe coal-fired plant requires some 2,300,000 tons of coal, versus some 175 tons of uranium (U_3O_8) from about 80,000 tons of 0.2 % ore for an equivalent nuclear powered plant (assuming a 65 percent plant factor).

Routine releases of radioactivity contained in the air and liquid effluents from nuclear power plants were the subject of a bitter controversy a couple of years ago in the United States; however, it is rarely heard of now. A principal argument advanced earlier was that, while the dose to the public from routine reactor operations has been extremely low and has constituted only a very small fraction of the so-called "standards" for such releases, the "as low as practicable" philosophy had not been explicitly written into the

AEC regulations. Critics maintained therefore that even though releases were much lower than the "standards", industry had been notorious in releasing up to the absolute limit allowed by law so that this would eventually happen with nuclear reactors. (They implied further that the 0.17 rem/yr. standard applicable to the public at large could be reached from the reactor emissions alone, which I shall show is impossible.) This was proclaimed in the face of the fact that releases from nuclear reactors had been maintained at an extremely small percentage of the "standards" long before the radiation controversy became a popular subject in the press. Nonetheless, the Atomic Energy Commission did propose to, in effect, lower the standards by a factor of close to 100, and to codify what always had been practiced, namely, that releases should be held to "as low as practicable". Even such vocal critics as Dr. Gofman now have stated flatly that routine releases represent no problem.

Effects of Radiation

In this connection, more should be said with respect to radiation and its effects. It has been recognized since almost from the time that Konrad Roentgen discovered x-rays in Würzburg, Germany, in 1895, that high doses of radiation are harmful. The carcinogenic potential of high doses of radiation were known within a few years of this discovery. This toxicity of radiation is not surprising, since almost all substances, including plain water can be severely damaging and even lethal if applied or taken in large enough quantities.

Radiation specialists pioneered a unique extension in toxicology, namely in considering what the effects of a toxic agent may be when the dose is progressively lowered. Historically, this problem had not been previously examined closely or quantitatively for most pharmacological or toxic agents. It was tacitly assumed that below some concentration or intake there were no effects, i. e., that a "threshold" dose existed. When one attempts to determine experimentally effects at progressively lower doses, at some point no effects can be found. This does not necessarily mean that there are none; it may simply mean that the percent affected in the exposed subjects is so small that one cannot practically deal with the total population required to find possible effects that are statistically significant. Radiation protectionists pioneered by extrapolating from high doses to very low doses, in order to estimate what might happen at these low doses. Because some data appear to be consistent with linearity (but consistent with other curves also), and sometimes for convenience (in health physics it is relatively easy to deal with potential radiation effects if every pertinent effect is assumed to be directly proportional to dose), the "linear model" emerged and has been widely utilized.

This model predicts a very small effect at low doses; nonetheless if one multiplies an extremely small probability times an extremely large population such as the two hundred million people in the United States, one obtains large absolute numbers. For instance, the number of Americans killed per year in auto accidents amounts to only approximately 0.025% of the total population. Yet this small percent of two hundred million Americans represents the approximately fifty thousand per year killed in auto accidents. These numbers were associated with the so-called radiation "standard" for the population at large, taken as 0.17 rem per year.

Standard setting bodies such as the International Commission on Radiation Protection (ICRP) have published risk estimates for many years [12]. Its "upper limit" number associated with 0.17 rem to 200,000,000 people is of the order of 4,000 to 5,000 additional cancer deaths per year. Such standard-setting bodies point out strongly and repeatedly that the assumptions that go into these estimates are extremely conservative, and that factors such as diminished rate of effect at low dose rates have not been taken into account. They take pains to point out that such estimates as those above may very well exaggerate the risk at low doses by a large factor, and that the actual number could be zero.

As a result of the radiation controversy, the National Academy of Sciences of the United States was asked recently (in 1970) to constitute a committee similar to the BEAR (Biological Effects of Atomic Radiations) Committee that looked deeply into the effects of radiations during the "50's". This new group, the so-called BEIR (Biological Effects of Ionizing Radiations) Committee [11] has recently produced an exhaustive survey of radiation effects. Their overall conclusions with respect to risk do not differ substantially from those that appeared in ICRP publications in 1965 and 1969 [12]. The Committee considered that the most likely estimate, based on the linear extrapolation model, is of the order of 5,100 to 6,800 new cancer deaths per year from continual exposure of the population of the United States to 0.17 rem per year.

In this report [11], the claims of Drs. Gofman and Tamplin that the number should be much higher are reviewed in detail, and reasons are given why their claims are considered to be excessive. It is also pointed out in the report that the average dose to the population from routine releases from reactors, now and in the foreseeable future, is a very small percent of that from background radiation or from medical exposures to x-rays.

A similar exhaustive review of radiation effects has also recently been published by the United Nations [13]. The individuals on the UN Committee of course reviewed the same data that the BEIR Committee reviewed, and it is not surprising that their risk estimates for high doses and dose rates are similar. In order to be conservative, the BEIR Committee chose to extrapolate (or more accurately, interpolate) from data obtained at high doses and dose rates to develop risk estimates at low doses and dose rates. The UN Committee considered this to be totally inadmissable. Thus the risk estimates they give for leukemia and other types of cancers are stated clearly to pertain only to the high dose levels at which data exist, and are not to be used for extrapolation to low doses and dose rates. They also point out that such extrapolation most likely would overestimate, propably by a considerable margin, the effects at low doses and dose rates.

It should be pointed out forcefully, that the association by any committee or individual of any number of cancer deaths with the 0.17 rem/yr. exposure of the entire population, has no relevance in the context of evaluating the risk of exposure to the routine releases from reactors. In the first place, the population exposure limit of 0.17 rem/yr. is *not* the limiting standard for these releases, and appears nowhere in the AEC regulatory documents pertaining to this subject. The primary standard that has been and is applied is a much more restrictive one, since it is based on the dose to the *individual* at the boundary of the site of the nuclear power plant (at the "fence post"). This figure for the individual was initially 0.5 rem per year; which has recently in effect been revised downward by the AEC by a factor of approximately 100. The current and proposed new "fence post" standards are enormously more restrictive than the 0.17 rem *average* dose, since by simple physical laws of dispersion and radioactive dacay, if the individual of the site boundary is allowed to receive even the 0.5 rem per year, the *average* dose to the entire population must be very much smaller than this [14, 15].

In Figure 5 is shown the rate of decrease in dose as a function of distance from a reactor. Notice that the dose rate falls by a factor of approximately 1,000 over a radius of 50 miles. From this we see that realistic assessments of the average dose to the population is smaller than the 0.17 rem per year by an extremely large factor for the past, present, and in the foreseeable future. Values depend on how conservative one wishes to make his estimates, and estimates of present exposure in the USA vary from the order of a tenth of a millirem per year down to a thousandth of a millirem per year, or less. Thus it is now generally accepted on this basis alone, that rather than the 32,000 or so cancer deaths that have been proclaimed widely in the context of nuclear power plants, the actual figure is more likely less than even one additional cancer death per year in the United States from all reactors now in operation or proposed.

A number of attempts have been made to use extensive population mortality statistics in a large country such as the USA, to see if a correlation exists between cancer or other deaths and the background radiation to which everyone is exposed [16]. Studies of this nature have shown a negative correlation, i. e., the cancer death rate was lower in high-background regions than in low-background regions. Such studies were fraught with statistical difficulties, and no firm conclusions have been possible.

Others have stated that a serious hazard, in addition to that from direct exposure to reactor effluents, exists because of substances such as radioiodine, cesium or strontium that may be released from the stack, deposited on the ground and vegetation, and in turn concentrated in the food chain,

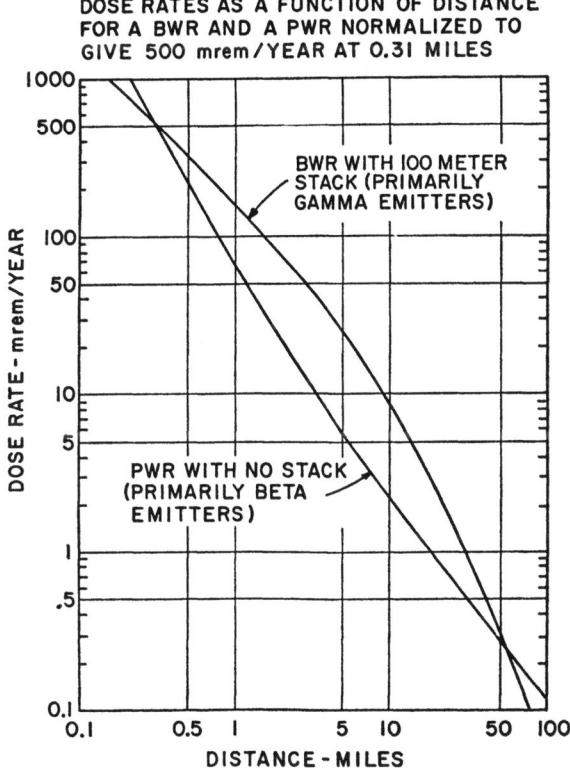

Fig. 5: Dose rates as a function of distance from light-water reactors. The curve is normalized to 500 millirem per year at the site boundary [14]

thus gaining access to man in larger quantities than by direct inhalation or ingestion. Extensive radiological surveillance has been done by the Bureau of Radiological Health (BRH) of the U. S. Department of Health, Education and Welfare, in the vicinity of a number of reactors. These detailed environmental studies included the Dresden reactor in Chicago, which was a "bad actor" in the sense that it was fueled with stainless steel-clad fuel elements, no longer used, that leaked more fission products into the reactor system and in turn into the environment than do more modern Zircalloy-clad elements. Nonetheless, the BRH group was unable to find any radioactivity attributable to the operation of this earlier plant in such materials as rain water, soil, cabbage, grass, milk, deer, rabbit, and fish. The overall conclusion [17] was, "... exposure to the surrounding population through consumption of food and water from radionuclides released by Dresden was not measurable".

Dr. Ernest Sternglass has proclaimed widely that radiations from fallout, nuclear reactors and fuel reprocessing plants have increased infant mortality rates, even though the radioactivity in the environment above background levels was essentially undetectable. A number of detailed analyses of the data he employed, summarized by Hull and Shore [18], conclude that the data do not support his thesis. The analysis indicate that Dr. Sternglass has selected only those parts of the data that appear to support his thesis and has ignored the remainder which in many cases is contradictory to his thesis. When one uses all of the data available in a given situation it becomes apparent that Dr. Sternglass' conclusions are not supported by the evidence.

Severe Accident

There is at present a great deal of confusion with respect to the probability of "severe" accidents and the extent of the damage to the public and to property that might result from such accidents. As of January 1973 there has been a total of 750 reactor-years experience throughout the world with commercial plants producing electricity, i. e. the equivalent of one plant operating for 7½ centuries. While there have been "incidents" in these plants as in any industrial endeavor, nuclear or otherwise, there has never been an accident in which the release of radioactivity has been known to injure even plant personnel, let alone the general public. This is an enviable record and indicates that the probability of severe accident is quite low. On the other hand one cannot conclude from this limited expericene that accidents are not possible and that they will not occur in the future.

In assessing the potential impact of reactor accidents it is perhaps useful to approach it historically. One of the most widely cited documents on this subject is the so-called WASH-740 Report, the findings of a committee established in the mid-50's by the USAEC to investigate the theoretical worst-imaginable consequences of a major accident in a large nuclear power plant [19]. Using the extreme assumptions it was directed to utilize, this committee postulated consequences of an accident that are truly horrendous, with thousands dead and multimillion dollar damages. It is useful to review the conditions under which the report was developed, and the remarks of some of those who were on the Committee.

In 1956 the Joint Committee on Atomic Energy of the U. S. Congress took up the question of public liability insurance for power reactors. The insurance underwriters had put up a pool of 60 million dollars, and this was the maximum insurance a reactor operator could obtain privately. The questions in the minds of the utility executives and of the AEC was whether this sum was adequate, or whether some form of government backup should be provided. Thus the study committee was asked to consider the cost in lives and property of a maximum "incredible" accident under the most adverse circumstances, namely the worst imaginable situation in which everything that could imagineably go wrong did in fact go wrong *simultaneously*.

In talking with some members of the study committee, I have learned that they asked repeatedly for some model to explain how these extreme circumstances could possibly occur, since they could not imagine any under which such a series of events could happen at all, let alone simultaneously. They were told in essence to forget if and how it could happen, and simply assess the consequences if it did happen. Thus, rather than being an evaluation of reactor safety, the report is a statement as to what would happen if one-half of the total inventory of radioactive material in a reactor of a given size were dispersed over a highly populated area under the worst conceivable atmospheric and other conditions. As one member of the committee expressed it [20], "... I believe it (the report) must rank as one of the most misused, and most misquoted documents in the whole nuclear energy field". He states, "At the outset, it is important to note that the WASH-740 has nothing to say about the safety of power reactors.... I repeat – the so-called 'Brookhaven-Report' is not concerned with reactor safety. Nonetheless, the document has been widely quoted by critics as constituting positive proof that not only can such a phenomenon and the results postulated in this study happen, but that the likelihood is appreciable that they will happen."

The authors assumed that every safeguard of the hypothetical reactor failed simultaneously, that its containment was breached at once and completely, and that atmospheric dispersion conditions were the poorest imaginable at the time of release. It was also assumed that the entire one half of the inventory of fission products was released as an aerosol, which is simply contrary to what is likely. It is extremely difficult to produce an aerosol of this material even if one wishes, let alone having it occur under the conditions of a reactor melt-down. The vast bulk of the material that might be hypothetically released from the fuel elements, even if it got through the fuel cladding, reactor vessel, biological shield, etc., would condense almost at once and could not be dispersed. In a real melt-down situation the reactor containment would not be breached for many minutes to an hour. Even under the worst conditions, it is highly probable that only a small part of it would be destroyed. Thus there would be time to use foam or other materials to retard releases from the containment. Not only was the postulated chain of events truly incredible, but the off-site consequences were vastly overestimated in WASH-740. The overall report is, as expressed by Dr. Kuper [20], a "mathematical exercise" rather than a scientific prediction. I might add parenthetically that, because of the perfect safety record of commercial power reactors, the insurance pool has been increased from sixty million dollars, to seventy-five million dollars, and now to eighty-two million dollars.

As I have previously indicated, more recently a group in Boston, the "Union of Concerned Scientists", headed by Ian Forbes [5, 21], has attacked the AEC and its regulation of reactor safety, particularly in regard to the reliability of the emergency core cooling system (ECCS). On the basis that the probability of a catastrophic accident is too high and that a great deal more research and analyses are required, they have called for a moratorium on all nuclear power plants. They have brought up a number of questions, and the fact that some of the group are technical men with impressive credentials has given the overall report an air of credibility. It is clear that they have raised a number of questions on reactor safety that must be answered, and that some may have validity. It is equally clear, however, that a large number of the people knowledgeable in this field do not agree at all with the Union's overall conclusions. However, there have been only spotty rebuttals and no detailed overall reply to the charges made. Perhaps the situation is similar to that which existed when the critics concerned with presumed radiation hazards made their first extensive and well publicized attacks. It is relatively easy to make widescale broadside attacks, and to attract the press and thus arouse the public if the claim is that a large number of people, particularly babies, will be exposed to "deadly" radiations or other supposed perils. It requires many days and weeks of dedicated and concentrated effort to deal responsibly with such attacks. If a reputation is warranted and issued, it is usually a collection of facts and analyses that represent dull reading which is unattractive to the press. Thus the public often hears or reads of the charges, but rarely the rebuttals.

Nonetheless, rebuttals have been put forth which indicate some of the defects in the Union of Concerned Scientists' reports [22, 23]. Bray [22] stated that the authors of the report should address themselves to certain questions which to my knowledge they have not yet done. These include: 1) they dwelt only on negative aspects of the situation, with no apparent knowledge and certainly no reference to sources identifying positive aspects. 2) They quoted only from an appendix of a U.S. Atomic Energy Commission report on ECCS, in which a particular situation was divorced from overall reactor safeguards and its consequences analyzed. The results are represented as a conclusion of the report. The fundamental conclusion of the overall report gave implications opposite to those from the appendix used by Forbes, et al. 3) The authors stated and implied that there is a serious lack of knowledge in a number of areas, when in fact there are numerous documents available which were not referred to or quoted by Forbes. Perhaps most serious is the fact that the authors used the consequences of WASH-740, which was not relevant, and used the consequences of the hypothetical accident put forth in their overall assessment of the result of ECCS's failure. As indicated above WASH-740 is not an authoritative reference to establish either the probability or the consequences of a core melt-down in a modern power reactor, with its many safeguard systems. 4) The authors implied that only one test was made of the ECCS, when in fact considerable test data are available. 5) The authors referred to an old report by C. G. Lawson, which flagged many technical issues that have been resolved in other more recent public documents readily available but not referred to by the authors. Bray concluded that "as a result, the contents of the article, under scrutiny, raise serious doubts as to its credibility" [22].

Rasmussen [23] has also expressed objections to the approaches and conclusions of Forbes, et al., and concludes that "Unfortunately, in this case, I believe that the critics have become so enmeshed in the details of one specific problem that they have lost the perspective of its overall implications. As a result, they have reached some conclusions that I believe are not in the best interest of the public they are trying to serve. I believe that the risks to the public from the operation of nuclear power stations of the type being built today are not significant when compared to the risks imposed by other activities of our highly technical society."

In no way do I wish to interfere with the activities or beliefs of any group, and certainly the UCS members can and should express their opinions. However, as I indicated in my introduction, one should be aware of the background of any individual or group who speaks or writes on these controversial issues. The authors of the critical reports indicated above are members of the Union of Concerned Scientists, a Boston area coalition of scientists, engineers and other professionals. Dr. Forbes is a nuclear engineer, Dr. Ford is an economist, Dr. Kendall is a nuclear and high energy physicist, and Dr. McKenzie is a nuclear physicist. None have had direct experience with reactor safety nor with the ECCS systems. The Union of Concerned Scientists was formed in March 1969, and has been most active in the areas of arms control and environmental pollution. It is an advocate organization concerned with environmental pollution from various sources. Most members of the UCS are also members of the Federation of American Scientists (FAS), which has been concerned over a number of years with the implications of scientific and technical developments on society. The UCS has made statements that imply an endorsement of their position by the FAS. The FAS has studied their position and has declined to endorse fully their views on the adequacy of reactor safety.

In a recent New York Times article (December 25, 1972) the UCS group dealt with the hazards of ship deliveries of liquefied natural gas. They stated that "an accident during ocean tanker transport near shore could release a cloud of combustible that would hug the earth long enough to be ignited in a nearby coastal city", with disasterous consequences. It is thus apparent that this group is oriented toward the extremes of potential hazards to the public.

A number of estimates relate to the actual propability of various reactor components failing and to the probability of the so-called "catastrophic accident". While there is some basis for such estimates of probability from a general knowledge of engineered facilities and their component parts, there is no "scientific" manner of arriving at such estimates with any degree of confidence. Weinberg [24] has referred to estimates of such extremely unlikely events as "trans-science". Thus such estimates are essentially a matter of judgment and educated guesses. True estimates of the probability of an occurrence can be obtained only when one has accumulated a large library of such events, to which one can apply statistical methods for predictive purposes. Since the number of significant power reactor incidents is zero, such an approach is not available. All such educated guesses indicate that the probability of serious malfunction and consequences are very low indeed. One estimate [25] is that the probability that the ECCS of a reactor will be called upon is perhaps less than once every ten thousand years of reactor operation, or of the order of 10^{-4} to 10^{-6} per reactor year. The probability that it will fail if called upon is an additional 10^{-2} to 10^{-4}. This gives a net probability of some release of activity into the containment, or even beyond it, to the range of 10^{-6} to 10^{-10} per reactor year.

In order to give some perspective as to the meaning of such probabilities, compare the risk figures given in Table III. Here are given the probabilities of various events, or the probability of an individual being seriously injured or killed in various situations. Notice that these range from about one in a hundred (10^{-2}) per year for serious injury in an automobile accident, to the order of one in ten million (10^{-7}) per year or so from living in the vicinity of a nuclear reactor. Further perspective can be gained by considering the reactions of society to events with different probabilities of occurrence [25]. With the exception of injury in a motor vehicle accident, it is difficult to find serious hazards from human activities with a probability of the order of 10^{-3} per person year or less. When a risk is this great, the individual or society usually takes action to reduce the hazard. This level of risk appears to be essentially unacceptable to everyone.

At the level of risk of 10^{-4} per person year, people are willing to spend money, especially public money, to control a hazard. For instance, traffic signs and control are provided, and fire departments are maintained. Campaigns are mounted to make people more aware of the risk, and there is an element of fear, e. g. "the life you save may be your own".

At risk levels of the order of 10^{-5} per person year, the hazards are still recognized and people are warned of the hazards, e. g. drowning, fire arms, poisoning, etc. Some people may accept some inconvenience to avoid the risk, such as avoiding air travel.

Accidents that occur with the probability of the order of 10^{-6} per person per year do not appear to be of great concern to the average person who feels that they can't happen to him. Accidents with a probability of less than 10^{-6} appear to most people to be in the "never or can't happen" category.

The public is willing to accept one level of risk with one type of hazard, but very different levels with another type of hazard. For instance, automobile accidents represent a real killer in the United States as in other developed countries. Nonetheless, this very substantial risk is pretty much shrugged of and forgotten in trade for the convenience of the automobile. Reports of airplane accidents in which even more that 100 persons per accident have been killed are newsworthy, usually, for no more than one day, and are remarkable for their lack of impact on the public in general or their habits with respect to flying. The public appear to take a very different view with respect to accidents that they can at least imagine are under their control, versus those over which they feel they have no control. Thus, while the average person has really very little control over whether or not he is going to be in an automobile accident, he likes to think that he does. On the other hand, radiation emanating from any community source is regarded as something inflicted on him by someone else. Also, radiation is particularly susceptible to being used to stir up public fear. It is something that the public in general does not understand, and such phrases as "no amount is safe" and "produces leukemia and cancer" can be used, however much out of context, to instill fear if not terror. Thus many of the public, particularly after being exposed to such treatments of the problem, appear quite unwilling to even examine, much less accept, an objective and rational appraisal of radiation risks in comparison to some greater risks that they encounter in everyday life. Radiation doses received by the public from different sources are given in Table IV.

Statistics can of course be misleading and are easily misunderstood or misused. There is the story of the man who was told that the probability of one person having a bomb aboard an airplane is 10^{-3}, and of two persons having a bomb, 10^{-6}. He then always carried a bomb with him to reduce the chances of someone else having such a device to 10^{-6}. This of course makes no sense. And, the fact that we are exposed to one risk in no way justifies a cavalier approach to other risks. However, it is obvious from Figure 8, that it makes no sense, if one's interest is in saving lives, to attempt to deny oneself or society the benefits of nuclear power on the basis of possible hazards from radioactivity releases in the vicinity. To be logical, one should concentrate on the much greater risks associated with such items as auto accidents, fires, smoking, or even the radiation exposure received in diagnostic x-ray, which is orders of magnitude greater than that received from reactor emanations.

My own evaluation of the overall accident situation, which the many engineers I have talked to agree with, is that, while the last word has not been said, an accident in which a core melt-down occurs is an extremely improbably accident, in the range of probabilities far less than the "act of God" or natural disaster type of event. Even in the most unlikely event that a core melt-down would occur, the probability of other backup safeguards failing is very remote. It appears doubtful that other than for radioiodine and volatiles, the bulk of the molten core would be broadly disseminated. Thus the probability of release of significant amounts of radioactivity into the environment is small indeed, and the risk to the public accordingly vanishingly small. Because of extensive and redundant safety systems, because of rigid quality control, and because of the continuing process of evaluating and reassessment *, the damage to the public from reactor accidents is most likely less than that which would be encountered from using alternative energy sources, or from having significant energy shortage.

It has also been reported that other scientists, for instance some at the Oak Ridge National Laboratory, are concerned with reactor safety. This is certainly true. However, from my knowledge of what these individuals have said, their concern is primarily with the curtailment of research on reactor safety. I certainly do not disagree with this concern. These individuals prefer, as do I, evaluations which are experimentally obtained, rather than those which depend upon extrapolated or assumed parameters. I certainly do not get the impression, however, that these individuals feel that the public is jeopardized by postulated "catastrophic accidents", or that power reactors are any less safe than I have indicated above.

Waste heat

The possible effects of waste heat discharged into bodies of water must be looked at carefully, whether it derives from fossil or nuclear plants. If the plant is located on a large body of water such as a very big lake or the ocean, then, with reasonable precautions, it would appear that the impact, if any, is minimal. Under such circumstances it would make little sense to insist on costly and potentially unsightly cooling towers. However, if power plants are to be located on rivers or bodies of water such as relatively small lakes, then the problem must be examined carefully, particularly if several reactors are built or are scheduled to be built on the same river or lake. Extensive studies have been done on the effect of one reactor on a river (the Haddam Neck reactor on the Connecticut River). These detailed studies have revealed no significant damage from the waste heat. Fish do migrate up the river past the plant with apparently no inference, and there have been no significant changes in the aquatic life in the river. Under circumstances where a number of plants are to be built, the impact must be assessed very carefully. The impact of sustained significant temperature changes, or of intermittant temperature changes, has not been fully assessed with respect to aquatic organisms. If it appears that there will be significant damage,

---

* The AEC requires the reporting of any significant incident in any type of reactor, and distributes these reports widely. Thus potential sources of trouble are "flagged" and additional preventive measures are taken.

there is no alternative at present to cooling towers. Approaches to utilizing waste heat for agricultural or other purposes are being investigated.

The question of estuaries requires special considerations since these are the breeding grounds for a wide variety of pelagic and other fish. If serious interference with the breeding habits of the fish are brought about either by thermal changes or by actual entrainment of small fish in the cooling systems of plants, the overall impact on commercial and sport fishing could be appreciable. In my estimation, no significant damage to estuaries of the United States has resulted as yet. A sufficient number of studies must be carried out to insure that the aquatic life is not adversely affected to any appreciable extent.

Long-term Waste Disposal

Spent fuel rods, on reprocessing, yield large quantities of radioactive material that cannot be destroyed and therefore must be stored in a manner such that it cannot gain access to the biosphere. The material contains nuclides with half-lives of various lengths. Thus some of the radioactivity decays away essentially before it is stored, some will decay away in a matter of a few years to a few decades, and some will remain radioactive for thousands of years. Thus the effort is to provide assurance that the material either will "never" gain access to the biosphere, or if and when it does, it will have decayed away to the point, and/or be sufficiently dilute such that the activity per unit mass of soil is comparable to that encountered from natural radioactivity.

Several approaches have been discussed and tried. Initially some radioactive wastes were deposited in the ocean; however, this practice has been, I believe, abandoned completely. The two principal approaches discussed in the United States are storage in salt beds or salt domes under the earth, or "engineered storage" on the surface of the earth. Extensive studies were done at Lyons, Kansas in salt mines, and this initially was received favorably by most if not all. The obvious advantage is that these salt mines are dry on a geologic time scale, and therefore the probability of dispersal of radioactive materials stored there is small indeed. Some critics objected, however, on the basis that there were a number of wells drilled in the area, and that by this means water might gain access to the stored material. The intent had been to fill these abandoned wells to prevent the access of water. Although study continues on the use of such salt mines as storage areas, this approach to the actual storage of wastes in the United States has been abandoned, at least temporarily [26].

The current approach is to use "engineered storage" at or near the surface. With this approach, radioactivity is imbedded in polymer material or concrete and stored in facilities on or near the surface. The advantage of this approach is that the material can be observed continuously and retrieved. Thus at a later date the material can be stored in some other fashion if newer or better approaches are devised.

The question arises as to the amount of land required for storage. While the amount of radioactive waste is large in terms of curies, the actual volume and the amount of land required is relatively small. Thus the land area that must be committed to such storage in the foreseeable future is of the order of a few thousand acres; a small area compared to that available and committed for other purposes.

The point is frequently raised, that such long term storage represents a strong commitment not only to the present society and its descendents, but perhaps even a different society at some date in the future. This is of course true, and the problem must be approached carefully. In my opinion, the problem of long-term storage is being approached carefully and is receiving a great deal of study. I believe it has been handled well, that means are available to handle it well on a continuing basis, and that the storage of waste does not and is not likely to constitute any perceptible hazard to the present nor a possible future civilization. It is, however, an important problem and must continue to receive detailed study and attention.

Transportation of Wastes

Sizable amounts of radioactivity have been transported over American highways for a number of decades, without known harm to the public. Spills have of course occurred; however, these have represented relatively small amounts of radioactivity that constitute more of a nuisance and a cleanup problem than any threat to public health. Requirements for the strength of casks for transportation have been made increasingly stringent, and now radioactive wastes are stored in containers that are designed to withstand the impact involved even in severe accidents. It must be remembered that such shipments of radioactivity are very clearly marked; there is no chance of an explosion of the cargo sufficiently severe to spread the material over large areas; and the confinement of any radioactivity that might be freed to the immediate environment will render its potential for public health hazard small indeed. Thus, while the nuisance value of a spill would be great, the hazard would be small.

The problem of transport of high level radioactive waste is an important one that must continue to receive attention and study. Preferably, fuel disposal areas should be in close juxtaposition to reprocessing plants, to minimize the extent of transportation of radioactive materials required. Nonetheless, in my opinion this problem is being handled well, and will be handled well in the future, with a correspondingly very small public health hazard involved. My personal preference by far, if I had to make a choice, would be to have many, many trains or trucks carrying high-level radioactive waste pass in the vicinity of my house, rather than trucks or railroad tank cars carrying gasoline, naphtalene or other known flammables. Transportation of these materials do frequently give rise to severe fires that cause real and not hypothetical deaths of people in the vicinity.

Fuel reprocessing Plants

Our experience with possible hazards from reprocessing plants for waste from LWR's in the United States indicates that this is very small [27, 28]. Essentially our only source of data is from the West Valley reprocessing plant in upper New York State. Here the dose from the emission of gaseous radionuclides is small. Radioactivity in liquids is released into a creek that flows through the area; however, only deer in the area, or perhaps a fisherman whose diet consists essentially only of fish from this creek would approach body burdens of radioactivity commensurate with the standards. There is of course always the potential for accident, with release of radioactivity. However, there is no possibility of anything like a "melt-down" that has been publicized as being possible in a reactor. Thus any accident would be on a relatively very small scale, with the consequences probably confined to the building in which it occurred or to the overall site. It is difficult to conceive of serious off-site consequences.

In Table V are given calculated release rates from the New York reprocessing plants for the year 1971 [28]. Very conservative assumptions were used in the calculations. Although the indicated "boundary" doses are higher than from some reactors, there will be far fewer reprocessing plants than reactors.

A Comparison of Reactor Types

Above I have dealt essentially entirely with light water reactors (LWR) of the boiling water reactor (BWR) and the pressurized water reactor (PWR) types. Routine emissions of radioactivity from both plants are extremely small. The airborne emission rate from PWR's seems to be smaller than that form BWR's, but their releases of tritium in liquid effluents are greater. The potential for a severe accident is extremely small in each, and I am not prepared to say whether the risk of accident with one is greater than that of the other.

Although gas-cooled reactors date back to the earliest days (the Oak Ridge X-10 Reactor; the Brookhaven Graphite Research Reactor; the Windscale Reactor in England) these were not power plants. They were air-cooled and used metallic uranium fuel. The operating temperatures were low. Considerable experience has been gained, principally in England, with intermediate temperature, closed circuit, gas-cooled reactors.

Commercial high temperature, gas-cooled reactors (HTGR or HTR) have been developed by private industry with government help, in the United States as in other countries. The first unit in the USA, Peach Bottom, was put into operation in Philadelphia in 1966 and has operated well. This was a small reactor rated at 40 MWe.

The first large HTGR in the United States has been constructed at Fort St. Vrain in Colorado, and it is expected to go into operation early this year. The fuel is UC_2 (92 % U-235 + Th-232 in the form of 100 to 300 micron pellets coated with carbon and SiC embedded in graphite). The moderator is graphite, and the coolant is high pressure helium gas. The reactor is rated at 330 MWe.

The advantages of HTGR's are the following: their thermal efficiency is higher than that of LWR's. It is in the order of 39.4 %, which is quite comparable to that of the best fossil plants. Routine emissions of radioactivity should be very low, although accurate data for large plants are not available. Extrapolating from the extremely low rates observed at the Peach Bottom plant, it is expected that the emissions from Fort St. Vrain reactor will be much less than one percent of the from LWR's of comparable power rating.

The high thermal efficiency of this plant has obvious advantages. Not only does it reduce the problem of waste heat and its disposal, it conserves water as well. Thus the plant is suitable for use in areas in which the available water is inadequate for cooling of less efficient types of energy generators.

The accident potential for the HTGR appears to be less than the already extremely small potential associated with light water reactors. Several factors contribute to this conclusion:

1) The possibility of uncontrolled power excursions is smaller because of the large negative temperature coefficient of the graphite moderator (the system responds very sluggishly to changes in reactivity).

2) The graphite has a large heat capacity and can withstand extremely high temperatures. Thus loss of coolant accidents does not produce violent temperature transients followed by the necessity of rapid insertion of emergency cooling to avoid core melt-down. In HTGR fission heat is absorbed by the enormous mass of graphite which rises in temperature very slowly. The design-base accident has been identified as "permanent loss of forced circulation" of primary coolant. This is an unlikely event which would require simultaneous failure of four independent redundant circulation systems. In the maximum credible accident, the calculated dose to any off-site resident is less than 1 rem.

Thus, although there has been less experience in the United States with HTGR's than in other countries, this type of reactor does seem to have definite advantages. Recently several orders have been placed for larger units, and I would expect to see the role of the HTGR in the overall energy production picture in the USA increase substantially over the next decade.

Breeder reactors are less developed in the United States than are those mentioned above. The obvious advantage of the breeder reactor is that it actually creates more fissionable fuel in the course of operation than it burns (conversion ratio greater than one), and thus the fuel supply becomes virtually inexhaustible. Most U. S. development effort has been expended on the liquid metal fast breeder reactor (LMFBR). The development program for the gas-cooled fast breeder reactor (GFBR) has received little federal support, but Gulf General Atomic has conducted extensive design studies.

Considerable experience has been accumulated with fast reactors and liquid sodium cooling. The technology for safe handling of large masses of liquid sodium seems to be well developed. Although the chain reaction with the fast breeder reactor has a smaller margin for control than do thermal reactors, the sophistication and reliability for modern control systems should adequately make up for this difference. The accident potential with the fast breeder reactor is much more difficult to assess than is that for LWR's or HTGR's. A number of factors have to be examined in more detail than they have been, and I am not prepared at this time to attempt any definitive statements on the subject.

Small test breeder reactors on the federal location in Idaho have been in operation for years, with no apparent significant difficulties. Although the Fermi reactor near Detroit was not a fast breeder, it had breeder characteristics. Developed jointly by the Government and private industry, Fermi early experienced serious difficulties due to a blocked coolant channel which took years to repair. It has recently run into severe financial problems, and is no longer in operation. President Nixon has stated that continued attention is to be given to the development of the LMFBR, and it is expected that a sizable demonstration plant will be developed over the next several years.

The routine emissions from breeder reactors should be extremely low, perhaps of the order of those from the HTGR. Although the breeder will of course produce plutonium, none will be released into the environment in the course of normal operations.

Plutonium Toxicity

The subject of plutonium toxicity merits a few comments. Plutonium has been widely alluded to by reactor critics and even some Congressmen as "the most toxic substance known to man". Any statement as extreme as this must of course be examined in some detail. There is no doubt that plutonium is capable of producing cancer, as are other radioactive materials, and the amount required to do this is small compared to most other isotopes. However, the plutonium must first be inhaled or ingested by an individual, in sufficient quantities, before any such effects are possible, and, barring the improbable accident, the plutonium in the reactor fuel is completely contained. One could equally well make such extreme statements with respect to the toxicity of a number of substances, e.g. botulinis toxin, some snake venoms, sulphuric acid, etc.; such statements are absolutely meaningless unless the route by which they are expected to gain access to the human body is spelled out.

During the forties, in Los Alamos, 27 men were exposed to plutonium via the inhalation route in the course of their work. All received appreciable amounts of plutonium, and of the 27 total, 20 received amounts equal to or greater than the present 0,04 μCi body burden standard for plutonium, Several received two to ten times this amount. It has been possible to follow 23 out of the 25 closely and they have been examined quite recently [29]. None of these individuals has developed cancer or other clinical symptoms attributable to the plutonium that they inhaled. The elapsed time is 27 years. The numbers are small and the statistics are correspondingly poor. Nonetheless, the data are at least sufficient to show that if plutonium is "the most toxic substance to man", it takes several decades for this toxicity to become manifest, even at relatively high levels commensurate with the body burden for radiation workers.

In the above statements I do not in any way wish to underplay the possible hazards of Pu exposure. It is a dangerous material that must not be released into the environment.

Risks of alternative Sources of Power

Rather recently, in the United States and elsewhere, individuals and groups have attempted to evaluate the problems and risks associated with sources of power other than nuclear, in order to come up with a basis for a realistic assessment of the role that each source should play under different conditions. These analyses are far from complete, and are sure to be extended and refined over the next few years.

One fact that strikes one immediately is that the risks associated with nuclear power have been investigated in greater detail than have those associated with any other power source or environmental pollutant. With respect to fossil fuels and other types of energy sources, relatively little attention has been paid to the possible hazards, either from the investigative or analytical standpoint. Data on the sources of distribution and hazards of effluents from the use of fossil fuels are being investigated, however, and thus the assessments will improve with time.

Perhaps one of the earliest approaches is that of Stig Bergstrom of Sweden, who compared the overall impact of fossil fuel versus nuclear power plants in Sweden [30]. His conclusion was that, "as to the choice between the oil and nuclear fuel alternatives the environmental impact from normal operations distinctly favors the nuclear plant". Similar studies have been carried out by Hull [31], who came to the same conclusion, that nuclear plants produce far less air pollution than their fossil cousins, and that, "... the catastrophic potential of nuclear plants has been vastly over-exaggerated by their adversaries". The consequences of coal mining operations in the United States have been detailed in a recent mineral industry survey [32]. The annual toll from coal mining far exceeds that from any other kind of mining, and in the single year 1971, 342 fatal injuries and approximately 24,000 non-fatal injuries occurred. The cost of pneumonoconiosis (black lung) disease from a total of approximately 350,000 claims between 1969 and 1971 amounted to approximately $ 530,000,000 or approximately $ 340,000,000 per year, about $ 2,100 per case.

The health effects of electricity generation from coal and nuclear fuels has been assessed by Lester Lave [33], with the conclusion that "the analysis results in an unequivocal conclusion that nuclear reactors offer less of a health hazard than do coal burning generators. Mining coal has roughly twelve times the accident rate as mining and milling uranium, per megawatt hour of electricity. Chronic disability associated with coal mining is about 26 times as great as that of uranium mining, per megawatt hour of electricity generated. The routine effluents of coal and nuclear generators have been examined. Both release comparable amounts of noble gases radiation during routine operation. Thus radioactive emissions are not a reason to prefer coal generators. Finally coal generators produce enormous amounts of air pollution which have been shown to be harmful to health.

Cohn and coworkers have studied the relationship between asthma and air pollution from a coal fueled power plant [34], and have come to the conclusion that "significant correlations were found between reported attack rate and temperature, and between attack rate and pollution levels after the effects of temperature had been removed from the analyses. These ... air pollution effects occurred at levels of pollution commonly found in large cities, and appeared greater at moderate than at low temperatures."

E. J. List looked carefully at energy requirements in California, and used as a "yardstick" Environmental Protection Agency (EPA) standards for air pollutants [35]. He came to the conclusion that "if we assume that all fuels (fossil) in these basins are burned at the minimum emission factors that appear technologically and economically feasible, the residual pollution is still such that the promulgated ambient air quality standards cannot be met. Thus, a simple policy of no-growth in these two basins still leaves the areas with significant air pollution".... continually lowering the emission factors will not attain clean air in the south coast and San Francisco area basins. ... the problem is aggravated by the increasing consumption of fuel every year. The only other policies available for air pollution control are either the relocation of energy demands to those energy sources with zero emission factors, or the curtailment of the use of fossil fuels as an energy source. The only near zero emission energy source capable of accomodating the possible demand for energy at this time is nuclear generated electric power. Hence ... the only way Los Angeles and San Francisco will attain ambient air quality satisfying promulgated standards is to replace fossil fuel consumption by nuclear power.

The results of such studies are shown in Table IX. Notice the enormous number of nuclear reactors that could be tolerated in the Los Angeles area, or the relatively few fossil fuel plants that could be permitted.

Hamilton [36] has made comparative studies, and has stated in effect that, if one uses the same very conservative assumptions in assessing the hazards from toxic agents in air polluted by the burning of fossil fuels, as one does in assessing the hazards of radiation, the numbers of injured and dead will be so large as to make any possible hazard from nuclear power production pale by comparison.

C. Starr and M. A. Greenfield have made an extensive study on the public health risks of thermal power plants [37]. This detailed study consists of a summary report plus five appendices. Both the situations of normal operation and potential accident situations are compared in detail. They use an approach that I have not seen developed in detail previously. In essence they state that, in assessing the hazards of any mode of energy production, one should use equally conservative assumptions throughout. Thus if one is going to use extremely conservative assumptions in assessing the probability of the consequences of a severe accident in a nuclear power plant, then the same extreme assumptions must be used in assessing the probability and results of accidents associated with the use of fossil fuel for power.

They also make the interesting observation that the relationship between federal standards for different pollutants, natural background levels of that

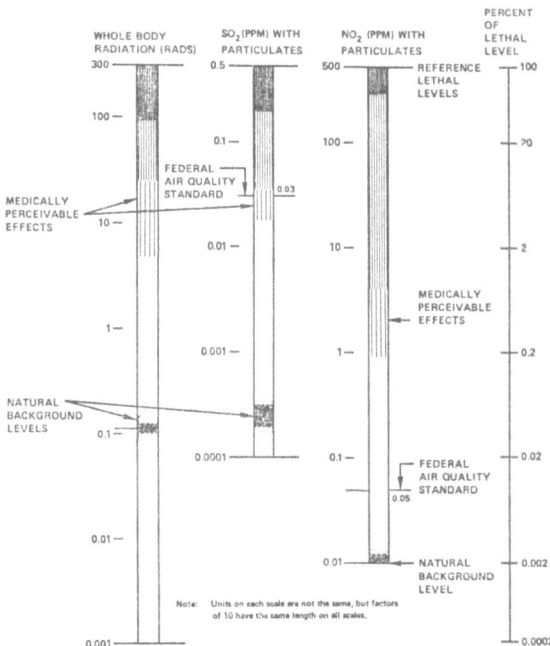

Fig. 6: Natural background levels, the level of the guidelines or standards, and levels at which medical effects are perceivable, for radiation, SO_2 and NO_2 [37]

same pollutant, and medically perceivable effects are vastly different for different pollutants (Figure 6). For instance, regulations or standards for radiation exposure of the public are at least two orders of magnitude lower than the levels of exposure at which one can begin to perceive medical effects. On the other hand, the federal air quality standard for SO_2 with particulates is set right at the level at which one can observe medical effects (Figure 7). It is ironic that even with a factor of 10^2 "cushion" in the radiation standards, there is still a great deal more concern about radiation exposure at levels even far below these already extremely conservative standards, than there is about the effects of SO_2 at or above the level of the standard set at the level where medical effects are readily perceived.

They have also pointed out that the severity of accidents in general varies inversely with their probability of occurrence. They have collected statistics on oil fires, and have compared these with the probability of accidental releases of radioactivity from PWR's. The results are shown in Figures 8 and 9. Note that the frequency of oil fires, real events, is much, much greater

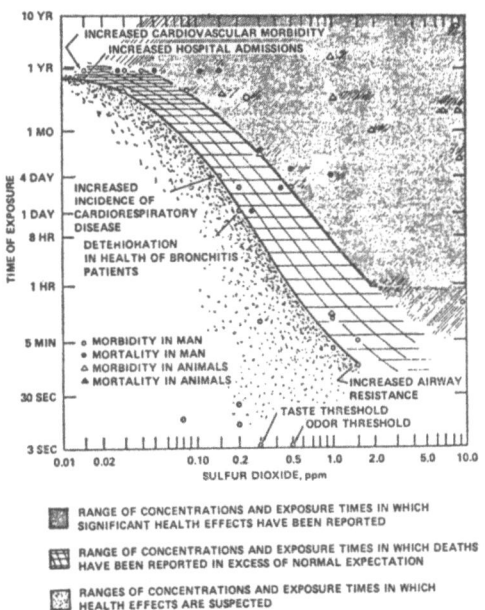

Fig. 7: The concentration of sulfur dioxide in the atmosphere, versus time of exposure, with indications of the effects on human beings following exposure to different concentrations over differing periods of time [37]. Taken from "Air Quality Criteria for Sulfur Dioxides", a Talk by Bernard E. Conley, Ph. D. Chief, Air Quality Criteria – National Center for Air Pollution Control.

than the postulated frequency of the release of significant quantities of radioactivity in a reactor accident. In Figure 10 they have translated these data into cumulative accident mortality from oil-fired versus nuclear plants, as a function of distance from the installation. At all distances, the impact of nuclear plants in terms of death is considerably lower than that from oil-fired plants.

Their overall conclusions are given in Figure 11 and Table VIII.

No one has done truly exhaustive studies on the possible catastrophic results of a large oil fire occurring in an urban community (such as in New

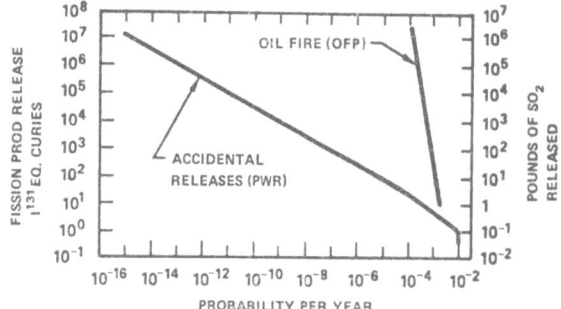

Fig. 8: A comparison of materials released following accidents involving pressure water reactors and oil plants [37]

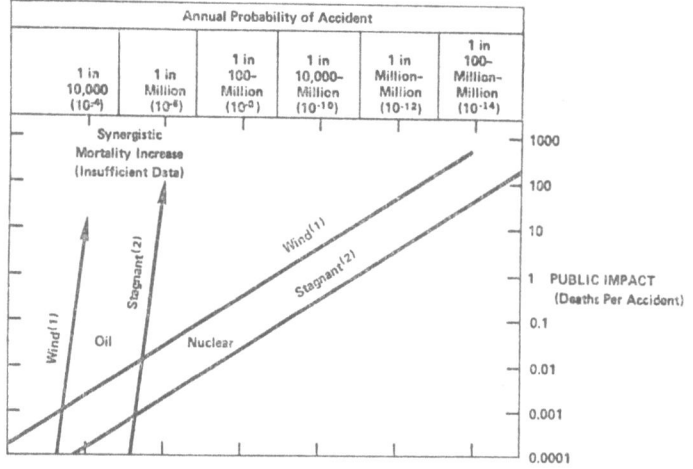

Fig. 9: A comparison of the public impact of individual accidents involving nuclear or oil powered power stations [37]

York where one large oil fire occurred very recently in the nearby community of Bayonne, New Jersey), making the extremely unrealistic and pessimistic assumptions used in a WASH-740 type of analysis. Under such circumstances one would postulate an extremely large fire, with the most adverse weather conditions such that the smoke spread out over a large urban area and was not dispersed by winds, and at a time when there was an extensive influenza epidemic. I am sure that if such an analysis were carried out in detail, the results would be horrendous to a degree comparable to those developed in the WASH-740 report.

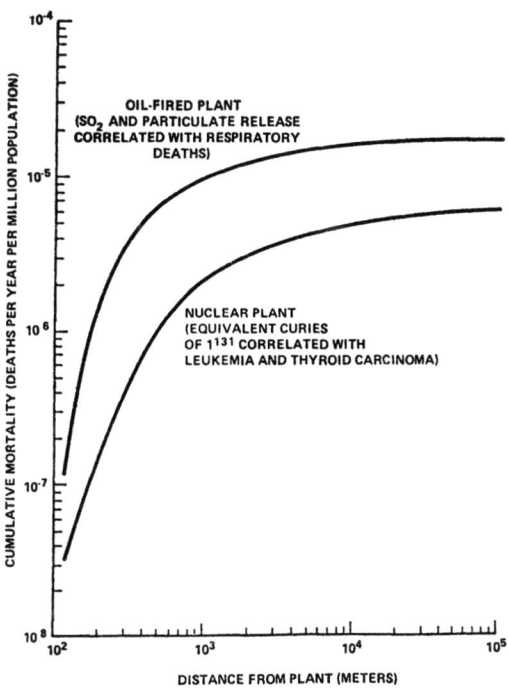

Fig. 10: A tabulation of mortality versus distance from the plant for nuclear fuel versus oil-fired plants [37]

Concluding Remarks

The "nuclear controversy" without question has had an enormous impact, not only on the power industry in the United States but on the development of all projects that might have an impact on the environment. It has resulted in a new general level of awareness, court decisions with far-reaching consequences, and new laws. The government and industry have become very concerned with their impact on the environment, and have taken positive steps to reduce the undesirable effects of their operations.

The overall tempo of the environmental movement has been reduced, but it has not and is not likely to disappear. Because of the broad awareness of the public that it has engendered, the basic concerns will persist and undoubtedly grow. The overall movement, although destructive in some instances, has probably had a net positive effect. Safety problems that did require more attention than they had received are now receiving that attention. Probably most important of all, the movement spurred an interest on the part of the public and the government not only in nuclear energy, but in the energy problems of the country in general. It is probable that, in the next few years, an overall energy policy will evolve. This most probably will take into account not only a realistic assessment of the energy needs of the country now and in the future, but the best ways of meeting these needs under different circumstances with the least impact on the environment. It seems inevitable that nuclear energy will play a very large role in satisfying the voracious energy needs of the country.

Summary

The American experience with respect to electrical power production has undergone and is undergoing considerable evolution. Early, frequently strident debates tended to focus on the risks and benefits of only one possible power source (nuclear), and on one possible hazard (radiation from routine releases). Discussions and analyses now have tended to become more moderate, and deal with not only one but all feasible power sources, as well as the costs and benefits associated with fuel production, the operation of such sources, fuel transportation and waste disposal. Radiation from routine releases is no longer a major issue. A number of attempts have been made to assess the total benefits and risks of nuclear versus fossil fuel plants, in order to allow a rational basis for deciding on what type of source is best for a given situation. More data and more refined analyses are needed; however, most of those completed to date have concluded that the overall cost in terms of possible damage to health and the environment is least for nuclear power, next for oil and the greatest for coal-fired plants. One very recent and extensive analytical report has concluded that the routine operation of a nuclear plant presents a smaller public health risk than routine operation of an oil-fired plant, by a factor of 10 to 100; and that the public health risk due to accidental releases from nuclear plants is less than that from oil-fired, by a factor of the order of 3. It is increasingly realized that there is and will continue to be a role for many energy resources in meeting society's needs, particularly during America's present and real "energy crisis". In addition to developing a long-lasting energy supply, such as through the nuclear approach, attention must be given to devising means of utilizing the limited fossil fuels, especially coal, in a manner that is much less contaminating and more efficient than at present. The net effect of the "controversy" has probably been positive, in that most appear to have found it necessary to moderate their positions to converge on positions that are more defensible and that are in the best interest of providing the energy needs of society with the least amount of potential hazard to the public and the environment.

Zusammenfassung

Die amerikanische Erfahrung hinsichtlich der Elektrizitätsproduktion hat bisher eine erhebliche Entwicklung durchgemacht und ist noch nicht abgeschlossen. Anfangs tendierten die häufig emotionellen Debatten vornehmlich auf die Risiken und Vorteile von seiten nur einer der möglichen Energiequellen (Kernenergie) und nur einer möglichen Gefährdung (reaktorproduzierte Strahlung). Heute neigen die Diskussionen und Analysen zu einer größeren Zurückhaltung und beschäftigen sich nicht nur mit einer, sondern zahlreichen möglichen Energiequellen und auch mit Kosten und Nutzen, die sich mit der Produktion der Brennstoffe, dem Betrieb solcher Quellen, dem Brennstofftransport und der Abfallbeseitigung ergeben. Strahlung vom Routinebetrieb der Kernkraftwerke ist nicht mehr eine besondere Streitfrage. In einer Reihe von Überlegungen sind die Gesamtvorteile und Risiken der Kernkraftwerke mit denen der konventionellen Kraftwerke verglichen worden, um rationell entscheiden zu können, welche Art von Energiequelle für eine gegebene Situation die beste ist. Mehr Erfahrung und sorgfältigere Analysen sind vonnöten; jedoch die meisten der bisher abgeschlossenen Untersuchungen stellen fest, daß der insgesamt zu veranschlagende Schaden für Gesundheit und Umwelt von seiten der Kernenergie niedriger ist als von seiten der mineralölnutzenden Werke und am größten für kohlebetriebene Kraftwerke ist. Ein vor kurzem veröffentlichter Bericht einer ausführlichen Untersuchung resultiert in der Feststellung, daß der Routinebetrieb eines Kernkraftwerks ein 10- bis 100mal kleineres Risiko für die öffentliche Gesundheit darstellt als der Routinebetrieb eines mineralölbetriebenen Werkes, und daß das Risiko für die Öffentlichkeit von seiten unfallbedingter Freisetzung von Radioaktivität von Kernkraftwerken etwa 3mal kleiner ist als das unfallbedingte Risiko von mineralölbetriebenen Werken. Es wird in zunehmendem Maße erkannt, daß gegenwärtig und auch in Zukunft viele Rohstoffe und Quellen für Energieproduktion zur Deckung des Bedarfs der Gesellschaft eine Rolle spielen – besonders während der augenblicklichen echten „Energiekrise" in den Vereinigten Staaten. Neben der Entwicklung langdauernder Energieversorgung, wie von seiten der Kernenergie, müssen Wege gefunden werden, um die begrenzten fossilen Brennstoffe, besonders

Kohle, in einer solchen Weise zu nutzen, die zu einer im Vergleich zur gegenwärtigen Situation geringeren Umweltverschmutzung und größeren Effizienz führt. Der Nettoeffekt der „nuklearen Kontroverse" war wahrscheinlich insofern positiv, als die meisten sich offensichtlich der Notwendigkeit bewußt wurden, ihre Positionen zu mäßigen bzw. solche Positionen anzustreben, die besser zu verteidigen sind und die im besten Interesse der Gesellschaft den Energiebedarf decken, mit dem geringsten Ausmaß an möglicher Gefahr für Öffentlichkeit und Umwelt.

References

[1] Hull, A. P., Reactor effluents: as low as practicable or as low as reasonable? Nuclear News, Vol. 15, No. 11, Nov. 1972.

[2] New York Times, 8 January 1973.

[3] Der Kölner Dom zerfällt zu Gips, X Magazin, January 1973.

[4] Air pollution across national boundaries. The impact on the environment of sulfur in air and precipitation. Sweden's case study for the U. N., Stockholm, Sweden, 1971.

[5] Forbes, I., Ford, D., Kendall, W., and MacKenzie, J., Nuclear reactor safety, an evaluation of new evidence. Nuclear news, September 1971.

[6] Sailor, V., Global energy needs. International Convocation, Manhattan College, New York, 14 October 1972.

[7] Sailor, V., Costs and benefits of nuclear power. New Horizons in Physics Lectures, State University College, New Paltz, New York, 17 October 1972.

[8] Newsweek, 22 January 1973.

[9] Duane Chapman, Timothy Tyrell and Timothy Mount. Electricity demand growth and the energy crisis. Science 178: 703-708, 1972.

[10] Biological Effects of Atomic Radiation Summary Reports. National Academy of Sciences/National Research Council, 1956.

[11] The effects on populations of exposure to low levels of ionizing radiation. Report of the Advisory Committee on the Biological Effects of Ionizing Radiations, Division of Med. Sc., National Academy of Sciences/National Research Council, November 1972.

[12] Radiosensitivity and spatial distribution of dose. ICRP Publication 14, Pergamon Press 1969.

[13] Ionizing Radiation: levels and effects. A report of the United Nations Scientific Committee on the Effects of Atomic Radiation to the General Assembly, No. E.72.IX.17, 1972.

[14] Rogers L. and Gamertsfelder, C., USA regulations for the control of releases of radioactivity into the environment in effluents from nuclear facilities. Environmental Aspects of Nuclear Power Stations. IAEA publication, 1971.

[15] Bond, V. P., Radiation standards, particularly as related to nuclear power plants. Health Physics Society Newsletter, 18 February 1971.

[16] Frigerio, N., Cancer epidemiology and the radiation background. Argonne National Laboratory, Argonne, Illinois. In press.

[17] Kahn, Bernd, et al., Radiological surveillance studies at a Boiling Water Nuclear Power Reactor. United States Public Health Service Report BRH/DER 70-1 March 1970.

[18] Hull, A. P., and Shore, F. J., Sternglass: A case history. Presented at the Atomic Industrial Forum Conference on Nuclear Public Interaction, Los Angeles, March 8, 1972.

[19] Theoretical possibilities and consequences of major accidents in large nuclear power plants. Report WASH-740, United States Atomic Energy Commission, Division of Technical Information, March 1957.

[20] Kuper, J. B. Horner, Testimony at the Shoreman Nuclear Power Station Hearings, Docket 50-322, May 25, 1970.

References

[21] Ford, D. F., Kendall, H. W., and MacKenzie, J. J., A critique of the new AEC design criteria for reactor safety systems. Union of Concerned Scientists, Cambridge, Mass., October 1971.

[22] Letter to the Editor, Nuclear News, November 1971, p 25.

[23] Rasmussen, Norman C., Nuclear reactor safety – An opinion. Nuclear News, January 1972, p 35.

[24] Weinberg, A. M., Social institutions and nuclear energy. Science 177: 27–34, 1973 (see also the editorial in the same issue).

[25] Otway, H. J., and Erdmann, C., Reactor siting and design from a risk viewpoint. Nuclear Engineering and Design 13: 365–376, 1970.

[26] Pittman, F., Management of commercial high-level waste. Speech, American Nuclear Society Meeting, Washington, D. C., 16 November 1972.

[27] Knox, J. B., Airborne radiation from the nuclear power industry. Nuclear News, February 1971.

[28] Martin, J. A., Calculations of environmental radiation exposures and population doses due to effluents from a nuclear fuel reprocessing plant. Environmental Protection Agency, Radiation Data and Reports, February 1973, Paper 59-75.

[29] Hempelmann, L., et al., A twenty-seven year study of selected Los Alamos plutonium workers. In press.

[30] Bergstrom, S., Environmental consequences from the normal operation of an urban nuclear power plant. Presented at the Health Physics Society Mid-year Topical Symposium, Ramada Inn, Idaho Falls, Idaho, November 3-6, 1970.

[31] Hull, A., Some comparisons of the environmental risks from nuclear and fossil fuel power plants. Nuclear Safety Vol. 12, No. 3, May-June 1971.

[32] Injury experience and work time in the solid mineral mining industry, 1970-71. Mineral Industry Surveys, U. S. Department of the Interior, Bureau of Mines, Washington, D. C., 20240. F. T. Moyer, author, 1972 (August).

[33] Lave, L., The health effects of electricity generation from coal, oil and nuclear fuel. Presented at the Sierra Club Conference of Environmental Effects on Electricity Generation, Johnson, Vermont, January 1972. Graduate School of Industrial Administration, Carnegie-Melon University Report WP-93-71-2, April 1972.

[34] Cohn, A. A., Bromberg, S., Buchley, R. W., Heiderscheit, C. T., and Shy, C. M., Asthma and air pollution from a coal-fueled power plant, AJPH 62, 1181 to 1188, 1972.

[35] List, E. J., Energy use in California: implications for the environment. Environmental Quality Laboratory, California Institute of Technology, EQL Report, 3 December 1971.

[36] Hamilton, L. J., Personal communication.

[37] Starr, C., and Greenfield, M. A., Public health risks of thermal power plants. UCLA-Eng-7242, May 1972, University of California, Los Angeles. Short version in Nuclear News, October 1972.

Tables

Table I: Sources of energy consumed in the United States in the Year 1971 [6, 7]

Energy source	Consumption (10^{15} Btu*)	Percent
Coal	12.560	18.2
Petroleum (crude oil)	27.940	40.5
Natural gas (liquids)	2.552	3.7
Natural gas (dry)	22.734	32.9
Hydropower	2.833	4.1
Nuclear	0.391	0.57
Wood	(∼0.8)	
Total	69.010	

* British thermal units

Table II: Principal approaches that have been suggested to obtain additional power

1. Drastically reduced or no additional power production
2. Now available and useable
   a) Hydroelectric
   d) Natural gas
   c) Oil
   d) Coal
   e) Nuclear
      LWR, HTGR (or HTR), Breeder
3. Feasible in selected regions of the earth
   a) Geothermal
   b) Solar
4. Possible in the future
   a) Fusion reactors
   b) Improved breeder
   c) Solar
5. Other
   a) Wind
   b) Tidal action
   c) Ocean currents, such as the Gulf Stream
6. Approaches to improving use of currently available sources
   a) Gasification of coal
   b) Combined cycles (systems that combine gas and steam turbines)
   c) MHD
   d) Fuel cells
   e) Hydrogen

Table III: Statistical chances of serious injury or death per year from a variety of causes or agents

Auto accident (disability)	1 chance in	100
Cancer, all types and causes	1 chance in	700
Auto death	1 chance in	4,000
Fire death	1 chance in	25,000
The "Pill", death	1 chance in	25,000
Drowning	1 chance in	30,000
Electrocution	1 chance in	200,000
Reactor Emanations; site boundary (5 to 10 mrem/yr.)	Less than 1 chance in	1,000,000
Average for population within 50 miles of reactor	Less than 1 chance in	10,000,000

Table IV: Radiation doses received by the public from a number of sources in 1970

	Ave. Dose millirem/yr.
Natural background	100–150
Diagnostic X-ray	50–150
The "standards"	170
Weapons testing	3
Jet travel, watches, color TV, etc.	1
Nuclear Power plants	less than .001

Table V: Doses to individual human beings in the vicinity of the nuclear reprocessing plant located at West Valley, New York 1971 [28]

Average whole body doses for an individual in the maximally exposed group

	Yearly dose
Tritium transpiration and inhalation	0.02 mrem
Krypton-85 air submersion	0.01 mrem
Deer meat ingestion	0.5 mrem
Fish meat ingestion	2.5 mrem
Creek bank occupancy	2.8 mrem
Total	5.83 mrem

Table VI: Air pollutants emitted from fossil-fuel steam electric plants, one billion pounds per year

Fuel	CO_2	CO	SO_2	NO_x	Particulates	Hydrocarbons	Aldehydes
(Billion pounds per year)							
Year 1969 (actual)							
Gas	439	0.001	0.002	1.36	0.052	0.139	0.010
Oil	272	0.0003	3.85	1.12	0.086	0.021	0.011
Coal	1660	0.297	31.8	6.02	7.83	0.090	0.001
Year 1977 (estimated)							
Gas	466	0.002	0.002	1.17	0.055	0.148	0.011
Oil	529	0.001	2.46	1.68	0.168	0.040	0.002
Coal	2140	0.383	11.9	7.32	6.81	0.115	0.002

Table VII: Numbers of power plants calculated to be tolerable under current practices in Los Angeles County, California * [35]

Plant type	Critical pollutant	Tolerable number of 1000 Mwe plants (exclusive of pollutants from other sources)
Oil	SO_2	10
Natural gas	NO_2	23
Nuclear reactor (LWR)	Radioactive gases	160,000

*Based on the following assumptions:

1. Unspecified mixture of radioactive isotopes released from nuclear plant. (Most restrictive assumption based on 1 millirem.)
2. Compliance with 0,5 % by weight sulfur content for oil.
3. Air volume of Los Angeles County was assumed to be 3165 km³ which implies a mean inversion height of 300 m.
4. Ventilation of this volume requires 1 day.
5. Effluent volume rate for 1000 Mwe reactor is taken as 0.5 x 10⁶ cfm which is an estimated upper limit.

Table VIII: Summary of the results of the analyses of Starr and Greenvield [37]

A study of oil-fired and nuclear power plants in an urban setting indicates the following, provided currently available technology is used to protect health and safety.

1. The public health risk from routine operations of electricity generating plants using nuclear fuel or oil is in the range of the very low hazards to which the public is exposed by uncontrollable events of nature such as being struck by lightning or bitten by a venomous animal or insect (about one death per year in a million population).

2. Routine operation of a nuclear plant presents a significantly smaller public health risk than the routine operation of an oil-fired plant, typically by a factor of 10 to 100.

3. The public health risk due to *accidental* releases from either a nuclear or oil-fired plant are both of the same magnitude, and about one hundred thousand times smaller than the risk from *routine* operation of the plants.

4. The maximum hypothetical accidents associated with either plant type are not likely to be sufficiently large to have a significant public health impact when compared to the normal incidence of disease.

5. Both oil fired and nuclear plant structures should be designed to meet the earthquake forces expected at a particular site, and a basis for such a design does exist.

6. The risk associated with transporting spent nuclear fuel can be made small enough so that the location of the associated fuel reprocessing installation is a separable factor in siting nuclear power plants.

Table IX: A comparison of the public risks from accidents involving nuclear reactors and oil-fired plants [37]

Plant type	Expected annual averages (Deaths per 10 million population per 1000 MWe plant per year)	
	Continuous operation at regulated exposure limits	Total risk from accidents
Nuclear reactor (cancer deaths)	1	Negligible (0.00006)
Oil fired plant (Respiratory deaths)	60	Negligible (0.0002)

*Veröffentlichungen
der Arbeitsgemeinschaft für Forschung des Landes Nordrhein-Westfalen
jetzt der Rheinisch-Westfälischen Akademie der Wissenschaften*

Neuerscheinungen 1970 bis 1974

Vorträge N Heft Nr.		NATUR-, INGENIEUR- UND WIRTSCHAFTSWISSENSCHAFTEN
199	*J. Herbert Hollomon,* Norman, Okl. *Stewart Blake,* Menlo Park, Kalifornien *Emanuel R. Piore,* New York *Wilhelm Krelle,* Bonn *David B. Hertz,* New York	Systems Management
200	*Michael F. Atiyah*	Vector Fields on Manifolds
201	*Jan Tinbergen,* Rotterdam	Optimale Produktionsstruktur und Forschungsrichtung
	Hans A. Havemann, Aachen	Neue Aspekte der Entwicklungsländerforschung
202	*Peter Mittelstaedt,* Köln	Lorentzinvariante Gravitationstheorie
203	*Heinz Wolff,* London	Bio-Medical Engineering
	Alexander Naumann, Aachen	Strömungsfragen der Medizin
204	*Fritz Schroter,* Neu-Ulm	Vorschläge für eine neue Fernsehbildsynthese
	Henricus P. J. Wijn, Eindhoven	Werkstoffe der Elektrotechnik
205	*Thomas Szabo,* Paris	Elektrische Organe und Elektrorezeption bei Fischen
	Franz Huber, Köln	Nervöse Grundlagen der akustischen Kommunikation bei Insekten
206	*Franz Broich,* Marl-Hüls	Probleme der Petrolchemie
207	*Franz Grosse-Brockhoff,* Düsseldorf	Elektrotherapie des Herzens (Eröffnungsfeier am 6. Mai 1970)
208	*Wolfgang Zerna,* Bochum	Bautechnische Probleme bei der Erstellung von Kernkraftwerken
	Otto Jungbluth, Bochum	Sandwichflächentragwerke im konstruktiven Ingenieurbau
209	*Erwin Gartner,* Köln	Die Vergasung von festen Brennstoffen – eine Zukunftsaufgabe für den westdeutschen Kohlenbergbau
	Rudolf Schulten, Aachen	Reaktoren zur Erzeugung von Wärme bei hohen Temperaturen
	Werner Peters, Essen	Entwicklung von Verfahren zur Kohlevergasung mit Prozeßwärme aus THT-Reaktoren
210	*Léon H. Dupriez,* Löwen	Währungsprobleme der EWG
	Wilhelm Krelle, Bonn	Die Ausnutzung eines gesamtwirtschaftlichen Prognosesystems für wirtschaftliche Entscheidungen
211	*Bernhard Rensch,* Münster	Probleme der Gedächtnisspuren
	Helmut Ruska†, Düsseldorf	Was kann der Biologe noch von der Elektronenmikroskopie erwarten?
212	*Franz Koenigsberger,* Manchester	Die Wechselwirkung zwischen Forschung und Konstruktion im Werkzeugmaschinenbau
	Rolf Hackstein, Aachen	Quantitative Analyse von Mensch-Maschine-Systemen
213	*Günter Schmölders,* Köln	Die öffentlichen Ausgaben als Elemente einer konjunkturpolitisch orientierten Haushaltsführung
	Erich Potthoff, Köln	Die Einheit der Unternehmensführung bei dezentralen Verantwortungsbereichen
214	*Martin Schmeißer,* Dortmund	Plasmachemie – ein aktuelles Teilgebiet der präparativen Chemie
	Gerhard Fritz, Karlsruhe	Bildung und Eigenschaften von Carbosilanen
215	*Charles Sadron,* Orléans	Die biologischen Makromoleküle
	Adolphe Pacault, Talence/Bordeaux	Einführung in eine phänomenologische Untersuchung der Evolution von Systemen

216	Werner Th. O. Forßmann, Düsseldorf	Moderne Knochenbruchbehandlung im allgemeinen Krankenhaus
	Carl-Heinz Fischer, Düsseldorf	Forschungsergebnisse und erste Erfahrungen mit einem neuen Kunststoff-Füllungsmaterial für die Zahnbehandlung
217	Lothar Jaenicke, Köln	Sexuallockstoffe im Pflanzenreich
218	Gerard P. Baerends, Groningen	Moderne Methoden und Ergebnisse der Verhaltensforschung bei Tieren
	Martin Lindauer, Frankfurt/M.	Orientierung der Bienen: Neue Erkenntnisse – neue Rätsel
219	Fritz Micheel, Münster	Reaktionen im flüssigen Fluorwasserstoff; Bildung von Kohlenwasserstoffen aus Kohlenhydraten
	Burchard Franck, Münster	Biosynthese biologisch aktiver Naturstoffe
220	Basil Joseph Asher Bard, London	Die Arbeit der National Research Development Corporation und ihre Beurteilung für den industriellen Fortschritt
	Walter Charles Marshall, Harwell	Die Umorientierung eines Kernforschungslaboratoriums
221	Günter Ecker, Bochum	Klassische Probleme der Gaselektronik in moderner Sicht
	Werner Rieder, Zürich	Plasma als Schaltmedium
222	Sven Effert, Aachen	Biomedizinische Technik
	Ludwig E. Feinendegen, Jülich	Nuklearmedizin im interdisziplinaren Feld der Großforschung
223	Peter A. Klaudy, Graz	Energieübertragung durch tiefstgekühlte, besonders supraleitende Kabel
	Theodor Wasserrab, Aachen	Elektrospeicherfahrzeuge
224	Karl Steimel, Frankfurt/M.	Spurgeführter Schnellverkehr – Schnellverkehr auf der Grundlage des Rad–Schiene-Systems
	Herbert Weh, Braunschweig	Berührungsfreie Fahrtechnik für Schnellbahnen
225	Hans-Jürgen Engell, Düsseldorf	Sonderfälle der Korrosion der Metalle
	Winfried Dahl, Aachen	Die mechanischen Eigenschaften der Stähle – wissenschaftliche Grundlagen und Forderungen der Praxis
226	Wilhelm Dettmering, Essen	Entwicklungsschritte zur Überschallverdichterstufe
	Friedrich Eichhorn, Aachen	Verfahrenstechnische Entwicklung der Schweißtechnik und ihre Bedeutung für die industrielle Fertigung
227	Pierre Jollès, Paris	From Lysozymes to Chitinases: Structural, Kinetic and Crystallographic Studies
	Hugo W. Knipping, Köln	Tuberkulosebekämpfung in Tropenländern
228	Emanuel Vogel, Köln	Hückel-Aromaten
229	Gaston Dupouy, Toulouse	Microscopie électronique sous haute tension
	Jacques Labeyrie, Gif-sur-Yvette	L'astronomie des hautes énergies
230	André Lichnerowicz, Paris	Mathématique, Structuralisme et Transdisciplinarité
231	Donato Palumbo, Brüssel	Die Thermonukleare Fusion – ihre Aussichten, Probleme und Fortschritte – innerhalb der Europäischen Gemeinschaft
232	Oswald Kubaschewski, Teddington (England)	Praktische Anwendung der metallchemischen Thermodynamik
	Bruno Predel, Münster	Thermodynamik und Aufbau von Legierungen – einige neuere Aspekte
233	Klaus Wagener, Jülich	Entwicklung der irdischen Atmosphäre durch die Evolution der Biosphäre
234	Eduard Mückenhausen, Bonn	Die Produktionskapazität der Böden der Erde
	Hermann Flohn, Bonn	Globale Energiebilanz und Klimaschwankungen
235	Bernhard Sann, Aachen	Die Senkung der Maschinenleistung bei Steigerung der Gewinnungsleistung und die Einsteuerung von Maschinen für die schälende Gewinnung von Steinkohle
	Lothar Freytag, Westfalia Lünen	Möglichkeiten der Verwirklichung von Forschungs- und Versuchsergebnissen in der Konstruktion von Maschinen für die schälende Kohlengewinnung
236	Werner Reichardt, Tübingen	Verhaltensstudie der musterinduzierten Flugorientierung an der Fliege *Musca domestica*
	Werner Nachtigall, Saarbrücken	Biophysik des Tierflugs
237	Henry C. J. H. Gelissen, Wassenaar (Niederlande)	Maßnahmen zur Förderung der regionalen Wirtschaft, gesehen im Blickfeld der EWG
	Horst Albach, Bonn	Kosten- und Ertragsanalyse der beruflichen Bildung
238	Victor Potter Bond, Upton, New York	The Impact of Nuclear Power on the Public: The American Experience

ABHANDLUNGEN

Band Nr.		
15	*Gerd Dicke, Krefeld*	Der Identitätsgedanke bei Feuerbach und Marx
16a	*Helmut Gipper, Bonn, und Hans Schwarz, Münster*	Bibliographisches Handbuch zur Sprachinhaltsforschung, Teil I. Schrifttum zur Sprachinhaltsforschung in alphabetischer Folge nach Verfassern – mit Besprechungen und Inhaltshinweisen (Erscheint in Lieferungen: bisher Bd. I, Lfg. 1–7; und Bd. II, Lfg. 8–16)
17	*Thea Buyken, Bonn*	Das römische Recht in den Constitutionen von Melfi
18	*Lee E. Farr, Brookhaven, Hugo Wilhelm Knipping, Köln, und William H. Lewis, New York*	Nuklearmedizin in der Klinik. Symposion in Köln und Jülich unter besonderer Berücksichtigung der Krebs- und Kreislaufkrankheiten
19	*Hans Schwippert †, Düsseldorf, Volker Aschoff, Aachen, u. a.*	Das Karl-Arnold-Haus. Haus der Wissenschaften der Arbeitsgemeinschaft für Forschung des Landes Nordrhein-Westfalen in Düsseldorf. Planungs- und Bauberichte (Herausgegeben von Leo Brandt †, Düsseldorf)
20	*Theodor Schieder, Köln*	Das deutsche Kaiserreich von 1871 als Nationalstaat
21	*Georg Schreiber †, Münster*	Der Bergbau in Geschichte, Ethos und Sakralkultur
22	*Max Braubach, Bonn*	Die Geheimdiplomatie des Prinzen Eugen von Savoyen
23	*Walter F. Schirmer, Bonn, und Ulrich Broich, Göttingen*	Studien zum literarischen Patronat im England des 12. Jahrhunderts
24	*Anton Moortgat, Berlin*	Tell Chuēra in Nordost-Syrien. Vorläufiger Bericht über die dritte Grabungskampagne 1960
25	*Margarete Newels, Bonn*	Poetica de Aristoteles traducida de latin. Ilustrada y comentada por Juan Pablo Martir Rizo (erste kritische Ausgabe des spanischen Textes)
26	*Vilho Niitemaa, Turku, Pentti Renvall, Helsinki, Erich Kunze, Helsinki, und Oscar Nikula, Åbo*	Finnland – gestern und heute
27	*Ahasver von Brandt, Heidelberg, Paul Johansen, Hamburg, Hans van Werveke, Gent, Kjell Kumlien, Stockholm, Hermann Kellenbenz, Köln*	Die Deutsche Hanse als Mittler zwischen Ost und West
28	*Hermann Conrad †, Gerd Kleinheyer, Thea Buyken und Martin Herold, Bonn*	Recht und Verfassung des Reiches in der Zeit Maria Theresias. Die Vorträge zum Unterricht des Erzherzogs Joseph im Natur- und Völkerrecht sowie im Deutschen Staats- und Lehnrecht
29	*Erich Dinkler, Heidelberg*	Das Apsismosaik von S. Apollinare in Classe
30	*Walther Hubatsch, Bonn, Bernhard Stasiewski, Bonn, Reinhard Wittram, Göttingen, Ludwig Petry, Mainz, und Erich Keyser, Marburg (Lahn)*	Deutsche Universitäten und Hochschulen im Osten
31	*Anton Moortgat, Berlin*	Tell Chuēra in Nordost-Syrien. Bericht über die vierte Grabungskampagne 1963
32	*Albrecht Dihle, Köln*	Umstrittene Daten. Untersuchungen zum Auftreten der Griechen am Roten Meer
33	*Heinrich Behnke und Klaus Kopfermann (Hrsgb.), Münster*	Festschrift zur Gedächtnisfeier für Karl Weierstraß 1815–1965
34	*Joh. Leo Weisgerber, Bonn*	Die Namen der Ubier
35	*Otto Sandrock, Bonn*	Zur ergänzenden Vertragsauslegung im materiellen und internationalen Schuldvertragsrecht. Methodologische Untersuchungen zur Rechtsquellenlehre im Schuldvertragsrecht
36	*Iselin Gundermann, Bonn*	Untersuchungen zum Gebetbüchlein der Herzogin Dorothea von Preußen
37	*Ulrich Eisenhardt, Bonn*	Die weltliche Gerichtsbarkeit der Offizialate in Köln, Bonn und Werl im 18. Jahrhundert

38	*Max Braubach, Bonn*	Bonner Professoren und Studenten in den Revolutionsjahren 1848/49
39	*Henning Bock (Bearb.), Berlin*	Adolf von Hildebrand Gesammelte Schriften zur Kunst
40	*Geo Widengren, Uppsala*	Der Feudalismus im alten Iran
41	*Albrecht Dihle, Köln*	Homer-Probleme
42	*Frank Reuter, Erlangen*	Funkmeß. Die Entwicklung und der Einsatz des RADAR-Verfahrens in Deutschland bis zum Ende des Zweiten Weltkrieges
43	*Otto Eißfeldt†, Halle, und Karl Heinrich Rengstorf (Hrsgb.), Münster*	Briefwechsel zwischen Franz Delitzsch und Wolf Wilhelm Graf Baudissin 1866–1890
44	*Reiner Haussherr, Bonn*	Michelangelos Kruzifixus für Vittoria Colonna. Bemerkungen zu Ikonographie und theologischer Deutung
45	*Gerd Kleinheyer, Regensburg*	Zur Rechtsgestalt von Akkusationsprozeß und peinlicher Frage im frühen 17. Jahrhundert. Ein Regensburger Anklageprozeß vor dem Reichshofrat. Anhang: Der Statt Regenspurg Peinliche Gerichtsordnung
46	*Heinrich Lausberg, Münster*	Das Sonett *Les Grenades* von Paul Valéry
47	*Jochen Schroder, Bonn*	Internationale Zuständigkeit. Entwurf eines Systems von Zuständigkeitsinteressen im zwischenstaatlichen Privatverfahrensrecht aufgrund rechtshistorischer, rechtsvergleichender und rechtspolitischer Betrachtungen
48	*Günther Stökl, Köln*	Testament und Siegel Ivans IV.
49	*Michael Weiers, Bonn*	Die Sprache der Moghol der Provinz Herat in Afghanistan
51	*Thea Buyken, Köln*	Die Constitutionen von Melfi und das Jus Francorum

Sonderreihe
PAPYROLOGICA COLONIENSIA

Vol. I
Aloys Kehl, Köln

Der Psalmenkommentar von Tura, Quaternio IX (Pap. Colon. Theol. 1)

Vol. II
Erich Lüddeckens, Würzburg
P. Angelicus Kropp O. P., Klausen
Alfred Hermann und Manfred Weber, Köln

Demotische und
Koptische Texte

Vol. III
Stephanie West, Oxford

The Ptolemaic Papyri of Homer

Vol. IV
Ursula Hagedorn und Dieter Hagedorn, Köln,
Louise C. Youtie und Herbert C. Youtie,
Ann Arbor

Das Archiv des Petaus (P. Petaus)

SONDERVERÖFFENTLICHUNGEN

Der Minister für Wissenschaft und Forschung des Landes Nordrhein-Westfalen – Landesamt für Forschung –

Jahrbuch 1963, 1964, 1965, 1966, 1967, 1968, 1969, 1970 und 1971/72 des Landesamtes für Forschung

Verzeichnisse sämtlicher Veröffentlichungen der Arbeitsgemeinschaft für Forschung des Landes Nordrhein-Westfalen, jetzt der Rheinisch-Westfälischen Akademie der Wissenschaften, können beim Westdeutschen Verlag GmbH, 567 Opladen, Postfach 1620, angefordert werden.

MIX
Papier aus verantwortungsvollen Quellen
Paper from responsible sources
FSC® C105338

If you have any concerns about our products,
you can contact us on
ProductSafety@springernature.com

In case Publisher is established outside the EU,
the EU authorized representative is:
**Springer Nature Customer Service Center GmbH
Europaplatz 3, 69115 Heidelberg, Germany**

Printed by Libri Plureos GmbH
in Hamburg, Germany